EVERYDAY EARTH AND SPACE SCIENCE MYSTERIES

STORIES FOR INQUIRY-BASED SCIENCE TEACHING

EVERYDAY EARTH AND SPACE SCIENCE MYSTERIES

STORIES FOR INQUIRY-BASED SCIENCE TEACHING

Richard Konicek-Moran, EdD
Professor Emeritus
University of Massachusetts
Amherst

Botanical illustrations by
Kathleen Konicek-Moran

National Science Teachers Association

National Science Teachers Association

Claire Reinburg, Director
Jennifer Horak, Managing Editor
Andrew Cooke, Senior Editor
Wendy Rubin, Associate Editor
Agnes Bannigan, Associate Editor
Amy America, Book Acquisitions Coordinator

ART AND DESIGN
Will Thomas Jr., Art Director
Rashad Muhammad, Designer, cover and interior design
Additional illustrations by D. W. Miller

PRINTING AND PRODUCTION
Catherine Lorrain, Director

NATIONAL SCIENCE TEACHERS ASSOCIATION
David L. Evans, Executive Director
David Beacom, Publisher

1840 Wilson Blvd., Arlington, VA 22201
www.nsta.org/store
For customer service inquiries, please call 800-277-5300.

Library of Congress Cataloging-in-Publication Data
Konicek-Moran, Richard.
 Everyday Earth and space science mysteries : stories for inquiry-based science teaching / by Richard Konicek-Moran.
    pages cm
 Includes bibliographical references and index.
 ISBN 978-1-936959-28-0
 1. Geology--Study and teaching. 2. Space sciences--Study and teaching. 3. Storytelling in education. 4. Inquiry-based learning. 5. Induction (Logic) in children. I. Title.
 GE40.K655 2013
 550--dc23
                    2013001545

Cataloging-in-Publication Data are available from the Library of Congress for the e-book.
e-LCCN: 2013012913

NGSS correlations for this book can be found at *www.nsta.org/publications/press/extras/mysteries-earth.aspx*.

CONTENTS

The Stories and Background Materials for Teachers

acknowledgments

I would like to dedicate these stories and materials to the dedicated and talented teachers in the Springfield Public Schools in Springfield, Massachusetts. They have been my inspiration to produce materials that work with city as well as rural children.

I would like to thank the following teachers, educators, and administrators who have helped me by field-testing the stories and ideas contained in this book over many years. These dedicated educators have helped me with their encouragement and constructive criticism:

Richard Haller
Jo Ann Hurley
Lore Knaus
Ron St. Amand
Renee Lodi
Deanna Suomala
Louise Breton
Ruth Chappel
Theresa Williamson

Third-grade team at Burgess Elementary in Sturbridge, Massachusetts

Second-grade team Burgess Elementary in Sturbridge, Massachusetts

Fifth-grade team at Burgess Elementary in Sturbridge, Massachusetts

Teachers at Millbury, Massachusetts Elementary Schools

Teachers and children at Pottinger Elementary School, Springfield, Massachusetts

All the administrators and science specialists in the Springfield, Massachusetts public schools, who are too numerous to mention individually

My thanks also go out to all of the teachers and students in my graduate and undergraduate classes who wrote stories and tried them in their classes as well as using my stories in their classes.

I will always be in the debt of my advisor at Columbia University, the late Professor Willard Jacobson who made it possible for me to find my place in teacher education at the university level.

I also wish to thank Skip Snow, Jeff Kline, Jean and Rick Seavey and all of the biologists in the Everglades National Park with whom I have had the pleasure of working for the past ten years for helping me to remember how to be a scientist again. And to the members of the interpretation groups in the Everglades National Park, at Shark Valley and Pine Island who helped me to realize again that it possible to help someone to look without telling them what to see and to help me to realize how important it is to guide people toward making emotional connections with our world.

My sincere thanks goes to Claire Reinburg of NSTA who had the faith in my work to publish the original book and the second and third volumes and is now taking a chance on a fourth; and to Andrew Cooke, my editor, who helps me through the crucial steps. In addition I thank my lovely, brilliant, and talented wife, Kathleen for her support, criticisms, illustrations, and draft editing.

Finally I would like to dedicate these words to all of the children out there who love the world they live in and to the teachers and parents who help them to make sense of that world through the study of science.

Preface

Earth and Space Sciences (ESS) range from the entire Earth into all of space—the universe and its galaxies. I have often thought that ESS should be the culminating science course in high schools and colleges since it has roots in all of the other branches of science. One cannot study the history of the Earth without incorporating the life sciences that laid down fossils. One cannot study the planets without physics and biology. As the Mars rover lays fresh tracks across the red planet, it looks for evidence of water and other signs that might signify the existence at one time or another of life, or at least its prerequisites. Consider this quote from the *Frameworks* document (NRC 2012):

> As a result, the majority of research in ESS is interdisciplinary in nature and falls under the categories of astrophysics, geophysics, geochemistry, and geobiology. However, the underlying traditional discipline of geology, involving the identification, analysis, and mapping of rocks, remains a cornerstone of ESS. (p. 169)

In these everyday stories you will find climatology, weather, decomposition, and astronomy. You get to go along on a geology trip with students up Bare Mountain. You will study the concepts of time, evaporation, air and air pressure, and probability. All of these stories correspond with the *scientific principles*, the *crosscutting concepts*, and the *core ideas* suggested and explained in the National Research Council's *A Framework for K–12 Science Education* (2012).

These stories are packaged in separate subject matter volumes so that those teachers who teach only one of the three areas covered in these books can use them more economically. However, it bears repeating that the crosscutting concepts meld together the various principles of science across all disciplines. It is difficult, if not impossible, to teach about any scientific concept in isolation. Science is an equal opportunity field of endeavor, incorporating not only the frameworks and theories of its various specialties, but also its own structure and history.

We hope that you will find these stories without endings a stimulating and provocative opening into the use of inquiry in your classrooms. Be sure to become acquainted with the stories in the other disciplinary volumes and endeavor to integrate all the scientific practices, crosscutting concepts, and core ideas that inquiry demands.

INTRODUCTION

Case studies on how to use the stories in the classroom

I would like to introduce you to one of the stories from the first volume of *Everyday Science Mysteries* (Konicek-Moran 2008) and then show how the story was used by two teachers, Teresa, a second-grade teacher, and Lore, a fifth-grade teacher. Then in the following chapters I will explain the philosophy and organization of the book before going to the stories and background material. Here is the story, "Where Are the Acorns?"

Where are the acorns?

Cheeks looked out from her nest of leaves, high in the oak tree above the Anderson family's backyard. It was early morning and the fog lay like a cotton quilt on the valley. Cheeks stretched her beautiful gray, furry body and looked about the nest. She felt the warm August morning air, fluffed up her big gray bushy tail and shook it. Cheeks was named by the Andersons since she always seemed to have her cheeks full of acorns as she wandered and scurried about the yard.

"I have work to do today!" she thought and imagined the fat acorns to be gathered and stored for the coming of the cold times.

Now the tough part for Cheeks was not gathering the fruits of the oak trees. There were plenty of trees and more than enough acorns for all of the gray squirrels who lived in the yard. No, the problem was finding them later on when the air was cold and the white stuff might be covering the lawn. Cheeks had a very good smeller and could sometimes smell the acorns she had buried earlier. But not always. She needed a way to remember where she had dug the holes and buried the acorns. Cheeks also had a very small memory and the yard

was very big. Remembering all of these holes she had dug was too much for her little brain.

The Sun had by now risen in the east and Cheeks scurried down the tree to begin gathering and eating. She also had to make herself fat so that she would be warm and not hungry on long cold days and nights when there might be little to eat.

"What to do ... what to do?" she thought as she wiggled and waved her tail. Then she saw it! A dark patch on the lawn. It was where the Sun did not shine. It had a shape and two ends. One end started where the tree trunk met the ground. The other end was lying on the ground a little ways from the trunk. "I know," she thought. "I'll bury my acorn out here in the yard, at the end of the dark shape and in the cold times, I'll just come back here and dig it up! Brilliant Cheeks," she thought to herself and began to gather and dig.

On the next day she tried another dark shape and did the same thing. Then she ran about for weeks and gathered acorns to put in the ground. She was set for the cold times for sure!

Months passed and the white stuff covered the ground and trees. Cheeks spent more time curled up in her home in the tree. Then one bright crisp morning, just as the Sun was lighting the sky, she looked down and saw the dark spots, brightly dark against the white ground. Suddenly she had a great appetite for a nice juicy acorn. "Oh yes," she thought. "It is time to get some of those acorns I buried at the tip of the dark shapes."

She scampered down the tree and raced across the yard to the tip of the dark shape. As she ran, she tossed little clumps of white stuff into the air, and they floated back onto the ground. "I'm so smart," she thought to herself. "I know just where the acorns are." She did seem to feel that she was a bit closer to the edge of the woods than she remembered, but her memory was small and she ignored the feelings. Then she reached the end of the dark shape and began to dig and dig and dig!

And she dug and she dug and she dug! Nothing! "Maybe I buried them a bit deeper," she thought, a bit out of breath. So she dug deeper and deeper and still, nothing. She tried digging at the tip of another of the dark shapes and again found nothing. "But I know I put them here," she cried. "Where could they be?" She was angry and confused. Did other squirrels dig them up? That was not fair. Did they just disappear? What about the dark shapes?

HOW TWO TEACHERS USED "WHERE ARE THE ACORNS?"

Teresa, a veteran second-grade teacher

Teresa usually begins the school year with a unit on fall and change. This year she looked at the National Science Education Standards (NSES) and decided that a unit on the sky and cyclic changes would be in order. Since shadows were something that the children often noticed and included in playground games (shadow tag), Teresa thought using the story of "Cheeks" the squirrel would be appropriate.

To begin, she felt that it was extremely important to know what the children already knew about the Sun and the shadows cast from objects. She wanted to know what kind of knowledge they shared with Cheeks and what kind of knowledge they had that the story's hero did not have. She arranged the children in a circle so that they could see one another and hear one another's comments. Teresa read the story to them, stopping along the way to see that they knew that Cheeks had made the decision on where to bury the acorns during the late summer and that the squirrel was looking for her buried food during the winter. She asked them to tell her what they thought they knew about the shadows that Cheeks had seen. She labeled a piece of chart paper, "Our best ideas so far." As they told her what they "knew," she recorded their statements in their own words:

"Shadows change every day."
"Shadows are longer in winter."
"Shadows are shorter in winter."
"Shadows get longer every day."
"Shadows get shorter every day."
"Shadows don't change at all."
"Shadows aren't out every day."
"Shadows move when you move."

She asked the students if it was okay to add a word or two to each of their statements so they could test them out. She turned their statements into questions and the list then looked like this:

"Do shadows change every day?"
"Are shadows longer in winter?"
"Are shadows shorter in winter?"
"Do shadows get longer every day?"
"Do shadows get shorter every day?"
"Do shadows change at all?"
"Are shadows out every day?"
"Do shadows move when you move?

Teresa focused the class on the questions that could help solve Cheeks's dilemma. The children picked "Are shadows longer or shorter in the winter?" and "Do shadows change at all?" The children were asked to make predictions based on their experiences. Some said that the shadows would get longer as we moved toward winter and some predicted the opposite. Even though there was a question as to whether they would change at all, they agreed unanimously that there would probably be some change over time. If they could get data to support that there was change, that question would be removed from the chart.

Now the class had to find a way to answer their questions and test predictions. Teresa helped them talk about fair tests and asked them how they might go about answering the questions. They agreed almost at once that they should measure the shadow of a tree each day and write it down and should use the same tree and measure the shadow every day at the same time. They weren't sure why time was important except that they said they wanted to make sure everything was fair. Even though data about all of the questions would be useful, Teresa thought that at this stage, looking for more than one type of data might be overwhelming for her children.

Teresa checked the terrain outside and realized that the shadows of most trees might get so long during the winter months that they would touch one of the buildings and become difficult to measure. That could be a learning experience but at the same time it would frustrate the children to have their investigation ruined after months of work. She decided to try to convince the children to use an artificial "tree" that was small enough to avoid our concern. To her surprise, there was no objection to substituting an artificial tree since, "If we measured that same tree every day, it would still be fair." She made a tree out of a dowel that was about 15 cm tall and the children insisted that they glue a triangle on the top to make it look more like a tree.

The class went outside as a group and chose a spot where the Sun shone without obstruction and took a measurement. Teresa was concerned that her students were not yet adept at using rulers and tape measures, so she had the children measure the length of the shadow from the base of the tree to its tip with a piece of yarn and then glued that yarn onto a wall chart above the date when the measurement was taken. The children were delighted with this.

For the first week, teams of three went out and took daily measurements. By the end of the week, Teresa noted that the day-to-day differences were so small that perhaps they should consider taking a measurement once a week. This worked much better, as the chart was less "busy" but still showed any important changes that might happen.

As the weeks progressed, it became evident that the shadow was indeed getting longer each week. Teresa talked with the students about what would make a shadow get longer and armed with flashlights, the children were able to make longer shadows of pencils by lowering the flashlight. The Sun must be getting lower too if this was the case, and this observation was added to the chart of questions. Later, Teresa wished that she had asked the children to keep individual science notebooks so that she could have been more aware of how each individual child was viewing the experiment.

The yarn chart showed the data clearly and the only question seemed to be, "How long will the shadow get?" Teresa revisited the Cheeks story and the children were able to point out that Cheeks's acorns were probably much closer to the tree than the winter shadows indicated. Teresa went on with another unit on fall changes and each week added another piece of yarn to the chart. She was relieved that she could carry on two science units at once and still capture the children's interest about the investigation each week after the measurement. After winter break, there was great excitement when the shadow began getting shorter. The shortening actually began at winter solstice around December 21 but the children were on break until after New Years Day. Now, the questions became "Will it keep getting shorter? For how long?" Winter passed and spring came and finally the end of the school year was approaching. Each week, the measurements were taken and each week a discussion was held on the meaning of the data. The chart was full of yarn strips and the pattern was obvious. The fall of last year had produced longer and longer shadow measurements until the New Year and then the shadows had begun to get shorter. "How short will they get?" and "Will they get down to nothing?" questions were added to the chart. During the last week of school, they talked about their conclusions and the children were convinced that the Sun was lower and cast longer shadows during the fall to winter time and that after the new year, the Sun got higher in the sky and made the shadows shorter. They were also aware that the seasons were changing and that the higher Sun seemed to mean warmer weather and trees producing leaves. The students were ready to think about seasonal changes in the sky and relating them to seasonal cycles. At least Teresa thought they were.

On the final meeting day in June, she asked her students what they thought the shadows would look like next September. After a great deal of thinking, they agreed that since the shadows were getting so short, that by next September, they would be gone or so short that they would be hard to measure. Oh my! The idea of a cycle had escaped them, and no wonder, since it hadn't really been discussed. The obvious extrapolation of the chart would indicate that the trend of shorter shadows would continue. Teresa knew that she would not have a chance to continue the investigation next September but she might talk to the third-grade team and see if they would at least carry it on for a few weeks so that the children could see the repeat of the previous September data. Then the students might be ready to think more about seasonal changes and certainly their experience would be useful in the upper grades where seasons and the reasons for seasons would become a curricular issue. Despite these shortcomings, it was a marvelous experience and the children were given a great opportunity to design an investigation and collect data to answer their questions about the squirrel story at a level appropriate to their development. Teresa felt that the children had an opportunity to carry out a long-term investigation, gather data, and come up with conclusions along the way about Cheek's dilemma. She felt also that the standard had been partially met or at least was in progress. She would talk with the third-grade team about that.

Lore (pronounced Laurie), a veteran fifth-grade teacher

In September while working in the school, I had gone to Lore's fifth-grade class for advice. I read students the Cheeks story and asked them at which grade they thought it would be most appropriate. They agreed that it would most likely fly best at second grade. It seemed, with their advice, that Teresa's decision to use it there was a good one.

However, about a week after Teresa began to use the story, I received a note from Lore, telling me that her students were asking her all sorts of questions about shadows, the Sun, and the seasons and asking if I could help. Despite their insistence that the story belonged in the second grade, the fifth graders were intrigued enough by the story to begin asking questions about shadows. We now had two classes interested in Cheeks's dilemma but at two different developmental levels. The fifth graders were asking questions about daily shadows, direction of shadows, and seasonal shadows, and they were asking, "Why is this happening?" Lore wanted to use an inquiry approach to help them find answers to their questions but needed help. Even though the Cheeks story had opened the door to their curiosity, we agreed that perhaps a story about a pirate burying treasure in the same way Cheeks had buried acorns might be better suited to the fifth-grade interests in the future.

Lore looked at the NSES for her grade level and saw that they called for observing and describing the Sun's location and movements and studying natural objects in the sky and their patterns of movement. But the students' questions, we felt, should lead the investigations. Lore was intrigued by the 5E approach to inquiry (*engage, elaborate, explore, explain, and evaluate*) and because the students were already "engaged," she added the "elaborate" phase to find out what her students already knew. (The five Es will be defined in context as this vignette evolves.) So, Lore started her next class asking the students what they "knew" about the shadows that Cheeks used and what caused them. The students stated:

"Shadows are long in the morning, short at midday, and longer again in the afternoon."

"There is no shadow at noon because the Sun is directly overhead."

"Shadows are in the same place every day so we can tell time by them."

"Shadows are shorter in the summer than in the winter."

"You can put a stick in the ground and tell time by its shadow."

Just as Teresa had done, Lore changed these statements to questions, and they entered the "exploration" phase of the 5E inquiry method.

Luckily, Lore's room opened out onto a grassy area that was always open to the Sun. The students made boards that were 30 cm² and drilled holes in the middle and put a toothpick in the hole. They attached paper to the boards and drew shadow lines every half hour on the paper. They brought them in each afternoon and discussed their results. There were many discussions about whether or not it made a difference where they placed their boards from day to day.

They were gathering so much data that it was becoming cumbersome. One student suggested that they use overhead transparencies to record shadow data and then overlay them to see what kind of changes occurred. Everyone agreed that it was a great idea.

Lore introduced the class to the *Old Farmer's Almanac* and the tables of sunsets, sunrises, and lengths of days. This led to an exciting activity one day that involved math. Lore asked them to look at the sunrise time and sunset time on one given day and to calculate the length of the daytime Sun hours. Calculations went on for a good 10 minutes and Lore asked each group to demonstrate how they had calculated the time to the class. There must have been at least six different methods used and most of them came up with a common answer. The students were amazed that so many different methods could produce the same answer. They also agreed that several of the methods were more efficient than others and finally agreed that using a 24-hour clock method was the easiest. Lore was ecstatic that they had created so many methods and was convinced that their understanding of time was enhanced by this revelation.

This also showed that children are capable of metacognition—thinking about their thinking. Research (Metz 1995) tells us that elementary students are not astute at thinking about the way they reason but that they can learn to do so through practice and encouragement. Metacognition is important if students are to engage in inquiry. They need to understand how they process information and how they learn. In this particular instance, Lore had the children explain how they came to their solution for the length-of-day problem so that they could be more aware of how they went about solving the challenge. Students can also learn about their thinking processes from peers who are more likely to be at the same developmental level. Discussions in small groups or as an entire class can provide opportunities for the teacher to probe for more depth in student explanations. The teacher can ask the students who explain their technique to be more specific about how they used their thought processes: dead ends as well as successes. Students can also learn more about their metacognitive processes by writing in their notebooks about how they thought through their problem and found a solution. Talking about their thinking or explaining their methods of problem solving in writing can lead to a better understanding of how they can use reasoning skills better in future situations.

I should mention here that Lore went on to teach other units in science while the students continued to gather their data. She would come back to the unit periodically for a day or two so the children could process their findings. After a few months, the students were ready to get some help in finding a model that explained their data. Lore gave them globes and clay so that they could place their observers at their latitude on the globe. They used flashlights to replicate their findings. Since all globes are automatically tilted at a 23.5-degree angle, it raised the question as to why globes were made that way. It was time for the "explanation" part of the lesson and Lore helped them to see how the tilt of the Earth could help them make sense of their experiences with the shadows and the Sun's apparent motion in the sky.

The students made posters explaining how the seasons could be explained by the tilt of the Earth and the Earth's revolution around the Sun each year. They had "evaluated" their understanding and "extended" it beyond their experience. It was, Lore agreed, a very successful "6E" experience. It had included the engage, elaborate, explore, explain, and evaluate phases, and the added extend phase.

references

Konicek-Moran, R. 2008. *Everyday science mysteries*. Arlington, VA: NSTA Press.

CHAPTER 1

THEORY BEHIND THE BOOK

We have all heard people refer to any activity that takes place in a science lesson as an "experiment." Actually, as science is taught today, true experiments are practically nonexistent. Experiments by definition test hypotheses, which are themselves virtually nonexistent in school science. A hypothesis, a necessary ingredient in any experiment, is a human creation developed by a person who has been immersed in a problem for a sufficient amount of time to feel the need to come up with a theory to explain events over which he or she has been puzzled.

However, it is quite common and proper for us to investigate our questions without proper hypotheses. Investigations can be carried out as "fair tests," which are possibly more appropriate for elementary classrooms, where children often lack the experience of creating a hypothesis in the true scientific mode. I recently asked a fourth-grade girl what a "fair experiment" was and she replied, "It's an experiment where the answer is the one I expected." We cannot assume even at the fourth-grade level students are comfortable with controlling variables; it needs repeating.

A hypothesis is more than a guess. It will most often contain an "if… then…" statement, such as, "**If** I put a thermometer in a mitten and the temperature stays the same, **then** perhaps the mitten did not produce heat." In school science, predictions should also be more than mere guesses or hunches, but rather based on experience and thoughtful consideration. Consistently asking children to give reasons for their predictions is a good way to help them see the difference between guessing and predicting.

Two elements are often missing in most school science curricula: sufficient *time* to puzzle over problems that have some *real-life applications*. It is much more likely that students will use a predetermined amount of time to "cover" an area of study—pond life, for example—with readings, demonstrations, and a field trip to a pond with an expert, topped off with individual or group reports on various pond animals and plants, complete with shoe box dioramas and giant posters. Or there may be a study of the solar system, with reports on facts about the planets and culminating with a class model of the solar system hung from the ceiling. These are naturally fun to do, but the issue is that there are seldom any real problems—nothing into which the students can sink their collective teeth into and use their minds to ponder, puzzle, hypothesize, and experiment.

You have certainly noticed that most science curricula have a series of "critical" activities in which students participate that supposedly lead to an understanding of a particular concept. In most cases, there is an assumption that students enter the study of a new unit with a common view or a common set of preconceptions about certain concepts and the activities will move the students closer to the accepted scientific view. This is a particularly dangerous assumption, since research shows that students enter into learning situations with a variety of preconceptions. These preconceptions are not only well ingrained in the students' minds but are exceptionally resistant to change. Going through the series of prescribed activities will have little meaning to students who have pre-conceptions that have little connection to the planned lessons, especially if the preconceptions are not recognized or addressed.

Bonnie Shapiro, in her book, *What Children Bring to Light* (1994), points out in indisputable detail how a well-meaning science teacher ran his students through a series of activities on the nature of light without knowing that the students in the class all shared the misconception that seeing any object originates in the eye of the viewer and not from the reflection of light from an object into the eye. The activities were, for all intents and purposes, wasted, although the students had "solved the teacher" to the extent that they were able to fill in the worksheets and pass the test at the end of the unit—all the while doubting the critical concept that light reflecting from object to eye was the paramount fact and meaning of the act of seeing. *Solving the teacher* means that the students have learned a teacher's mannerisms, techniques, speech patterns, and teaching methods to the point that they can predict exactly what the teacher wants, what pleases or annoys her, and how to perform so the teacher believes her students have learned and understood the concepts she attempted to teach.

In her monograph *Inventing Density* (1986), Eleanor Duckworth says, "The critical experiments themselves cannot impose their own meanings. One has to have done a major part of the work already. One has to have developed a network of ideas in which to imbed the experiments." This may be the most important quote in this book!

How does a teacher make sure students develop a network of ideas in which to imbed the class activities? How does the teacher uncover student misconceptions about the topic to be studied? I believe that this book can offer some answers to these questions and offer some suggestions for remedying the problems mentioned above.

WHaT IS INQUIry, anyway?

There is probably no one definition of "teaching for inquiry," but at this time the acknowledged authorities on this topic have to be the National Research Council (NRC) and the American Association for the Advancement of Science (AAAS). After all, they are respectively the authors of the *National Science Education Standards* (1996) and the *Benchmarks for Science Literacy* (1993), upon which most states have based their curriculum standards. For this reason, I will use their definition, which I will follow throughout the book. The NRC, in *Inquiry and the National Science Education Standards: A Guide for Teaching and Learning* (2000), says that for real inquiry to take place in the classroom, the following five essentials must occur:

- Learner engages in scientifically oriented questions.
- Learner gives priority to evidence in responding to questions.
- Learner formulates explanations from evidence.
- Learner connects explanations to scientific knowledge.
- Learner communicates and justifies explanations. (p. 29)

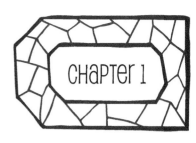
In essence the NRC strives to encourage more learner self-direction and less direction from the teacher as time goes on during the school years. The NRC also make it very clear that all science cannot be taught in this fashion. Science teaching that uses a variety of strategies is less apt to bore students and be more effective. Giving demonstrations, leading discussions, solving presented problems, and entering into a productive discourse about science are all viable alternatives. However, the NRC does suggest that certain common components should be shared by whichever instructional model is used:

- Students are involved with a scientific question, event, or phenomenon that connects with what they already know and creates a dissonance with their own ideas. In other words, they confront their preconceptions through an involvement with phenomena.
- Students have direct contact with materials, formulate hypotheses, test them, and create explanations for what they have found.
- Students analyze and interpret data, and come up with models and explanations from these data.
- Students apply their new knowledge to new situations.
- Students engage in metacognition, thinking about their thinking, and review what they have learned and how they have learned it.

You will find opportunities to do all of the above by using these stories as motivators for your students to engage in inquiry-based science learning.

THE reasons for THIS BOOK

According to a summary of current thinking in science education in the journal *Science Education,* "one result seems to be consistently demonstrated: students leave science classes with more positive attitudes about science (and their concepts of themselves as science participants) when they learn science through inductive, hands-on techniques in classrooms where they're encouraged by a caring adult and allowed to process the information they have learned with their peers" (1993).

This book, and particularly the stories that lie within, provide an opportunity for students to take ownership of their learning and as stated in the quotation above, learn science in a way that will give them a more positive attitude about science and to process their learning with their classmates and teachers. Used as intended, the stories will require group discussions, hands-on and minds-on techniques, and a caring adult.

THE STORIES

These stories are similar to mystery tales but purposely lack the final chapter where the clever sleuth finally solves the mystery and tells the readers not only "whodunit," but how she knew. Because of the design of the tales in this book, the students are challenged to become the sleuths and come up with likely "suspects" (the hypotheses or predictions) and carry out investigations (the experiments or investigations) to find out "whodunit" (the results). In other words, they write the final ending or perhaps endings. They are placed in a situation where they develop, from the beginning, "the network of ideas in which to imbed activities," as Duckworth suggests (1986, p. 39). The students are also the designers of the activities and therefore have invested themselves in finding the outcomes that make sense to them. I want them to have solved the problem rather than having solved the teacher. I do want to reemphasize, however, that we should all be aware that successful students do spend energy in solving their teachers.

In one story ("Cool It, Dude!"), Rosa and her friends wonder if crushed ice cools a drink faster than ice cubes. They have to consider the differences in heat transfer between the two shapes of the ice. Truly this is science as process and product. It also means that the students "own" the problem. This is what we mean by "hands-on, minds-on" science instruction. The teachers' belief in the ability of their students to own the questions and to carry out the experiments to reach conclusions, is paramount to the process. Each story has suggestions as to how the teachers can move from the story reading to the development of the problems, the development of the hypotheses and eventually the investigations that will help their students to come to conclusions.

Learning science through inquiry is a primary principle in education today. You might well ask, "instead of what?" Well, instead of learning science as a static or unchanging set of facts, ideas, and principles without any attention being paid to how these ideas and principles were developed. Obviously, we cannot expect our students to discover all of the current scientific models and concepts. We do however, expect them to appreciate the processes through which the principles are attained and verified. We also want them to see that science includes more than just what occurs in a classroom—that the everyday happenings of their lives are connected to science. Exploring evaporation, wondering about the phases of the Moon around the world, and examing frost on windows are only some of the examples of everyday life connected to science as a way of thinking and as a way of constructing new understandings about our world.

There are 19 stories in this book, each one focused on a particular conceptual area, such as astronomy, transfer of heat, evaporation, geology, and geography. Each story can be photocopied and distributed to students to read and discuss or they can be read aloud to students and discussed by the entire class. During the discussion, it is ultimately the role of the teacher to help the students to find the problem or problems and then design ways to find out answers to the questions they have raised.

Most stories also include a few "distractors," also known as common misconceptions or alternative conceptions. The distractors are usually placed in the stories as opinions voiced by the characters who discuss the problematic situation. For example, in "Here's the Crusher," family members ponder why a plastic bottle is collapsing. Each family member has his or her own preconception or misconception. The identification of these misconceptions is the product of years of research, and the literature documents the most common, often shared by both children and adults. Where do these common misconceptions come from and how do they arise?

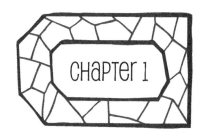

DEVELOPMENT OF MENTAL MODELS

Until recently, educational practice has operated under the impression that children and adults come to any new learning situation without the benefit of prior ideas connected to the new situation. Research has shown that in almost every circumstance, learners have developed models in their mind to explain many of the everyday experiences they have encountered (Bransford, Brown, and Cocking 1999; Watson and Konicek 1990; Osborne and Fryberg 1985). Everyone has had experience with differences in temperature as they place their hands on various objects. Everyone has seen objects in motion and certainly has been in motion, either in a car, plane, or bicycle. Everyone has experienced forces in action, upon objects or upon themselves. Finally, each of us has been seduced into developing a satisfactory way to explain these experiences and to have developed a mental model, which explains these happenings to our personal satisfaction. Probably, most individuals have read books, watched programs on TV or in movie theaters, and used these presented images and ideas to embellish their personal models. It is even more likely that they have been in classrooms where these ideas have been discussed by a teacher or by other students. The film *A Private Universe* (Schnepps 1987) documents that almost all of the interviewed graduates and faculty of Harvard University showed some misunderstanding for either the reasons for the seasons, or for the reasons for the phases of the Moon. Many had taken high-level science courses either in high school or at the university.

According to the dominant and current learning theory called *constructivism*, all of life's experiences are integrated into the person's mind; they are accepted or rejected or even modified to fit existing models residing in that person's mind. Then, these models are used and tested for their usefulness in predicting outcomes experienced in the environment. If a model works, it is accepted as a plausible explanation; if not, it is modified until it does fit the situations one experiences. Regardless, these models are present in everyone's minds and brought to consciousness when new ideas are encountered. They may be in tune with current scientific thinking but more often they are "common sense science" and not clearly consistent with current scientific beliefs.

One of the reasons for this is that scientific ideas are often counterintuitive to everyday thinking. For example, when you place your hand on a piece of metal in a room, it feels cool to your touch. When you place your hand on a piece of wood in the same room it feels warmer to the touch. Many people will deduce that the temperature of the metal is cooler than that of the wood. Yet, if the objects have been in the same room for any length of time, their temperatures will be equal.

It turns out that when you place your hand on the metal, it conducts heat out of your hand quickly, thus giving the impression that it is cold. The wood does not conduct heat as rapidly as the metal and therefore 'feels' warmer than the metal. In other words, our senses have fooled us into thinking that instead of everything in the room being at room temperature, the metal is cooler than anything else. Therefore our erroneous conclusion is that metal objects are always cooler than other objects in a room. Indeed, if you go from room to room and touch many objects, your idea is reinforced and becomes more and more resistant to change.

These ideas are called by many names: *misconceptions, prior conceptions, children's thinking,* or *common sense ideas.* They all have two things in common. They are usually firmly embedded in the mind and they are highly resistant to change. Finally, if allowed to remain unchallenged, these ideas will dominate a student's thinking, for example, about heat transfer, to the point that the scientific explanation will be rejected completely regardless of the method by which it is presented.

Our first impression is that these preconceptions are useless and must be quashed as quickly as possible. However, they are useful since they are the precursors of new thoughts and should be modified slowly toward the accepted scientific thinking. New ideas will replace old ideas only when the learner becomes dissatisfied with the old idea and realizes that a new idea works better than the old. It is our role to challenge these preconceptions and move learners to consider new ways of looking at their explanations and to seek ideas that work in broader contexts with more reliable results.

WHY STORIES?

Why stories? Primarily, stories are a very effective way to get someone's attention. Stories have been used since the beginning of recorded history and probably long before that. Myths, epics, oral histories, ballads, dances, and such have enabled humankind to pass on the culture of one generation to the next, and the next, *ad infinitum.* Anyone who has witnessed story time in classrooms, libraries, or at bedtime knows the magic held in a well-written, well-told tale. They have beginnings, middles, and ends.

These stories begin like many familiar tales do: in homes or classrooms; with children interacting with siblings, classmates, or friends; with parents or other adults in family situations. But here the resemblance ends between our stories and traditional ones.

Science stories normally have a theme or a scientific topic that unfolds giving a myriad of facts, principles, and perhaps a set of illustrations or photographs, which try to explain to a child the current understanding about the given topic. For years science books have been written as reviews of what science has constructed to the present. These books have their place in education, even though children often get the impression from these books that the information they have just read about appeared magically as scientists went about their work and "discovered"

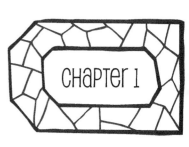

truths and facts depicted in those pages. But as Martin and Miller (1990) put it: "The scientist seeks more than isolated facts from nature. The scientist seeks a *story* [emphasis mine]. Inevitably the story is characterized by a *mystery.* [Emphasis mine]. Since the world does not yield its secrets easily, the scientist must be a careful and persistent observer."

As our tales unfold, discrepant events and unexpected results tickle the characters in the stories and stimulate their wonder centers making them ask, "What's going on here?" Most important of all, our stories have endings which are different than most. They are the mysteries that Martin and Miller talk about. They end with an invitation to explore and extend the story and to engage in inquiry.

These stories do not come with built-in experts who eventually solve the problem and expound on the solution. There is no Doctor Science who sets everybody straight in short order. Moms, dads, big sisters, brothers, and friends may offer opinionated suggestions ripe for consideration, or tests to be designed and carried out. It is the readers who are invited to become the scientists and solve the problem.

references

American Association for the Advancement of Science (AAAS).1993. *Benchmarks for science literacy.* New York: Oxford University Press.

Bransford, J. D., A. L. Brown, and R. R. Cocking, eds. 1999. *How people learn.* Washington, DC: National Academies Press.

Duckworth, E. 1986. *Inventing density.* Grand Forks, ND: Center for Teaching and Learning, University of North Dakota.

Martin, K., and E. Miller. 1990. Storytelling and science. In *Toward a whole language classroom: Articles from language arts, 1986–1989*, ed. B. Kiefer. Urbana, IL: National Council of Teachers of English.

National Research Council (NRC). 2000. *Inquiry and national science education standards: A guide for teaching and learning.* Washington, DC: National Academies Press.

Osborne, R., and P. Fryberg. 1985. *Learning in science: The implications of children's science.* Auckland, New Zealand: Heinemann.

Schneps, M. 1996. *The private universe project.* Harvard Smithsonian Center for Astrophysics.

Shapiro, B. 1994. *What children bring to light.* New York: Teachers College Press.

Watson, B., and R. Konicek. 1990. Teaching for conceptual change: Confronting children's experience. *Phi Delta Kappan* 71 (9): 680–684.

CHAPTER 2
USING THE BOOK AND THE STORIES

It is often difficult for overburdened teachers to develop lessons or activities that are compatible with the everyday life experiences of their students. A major premise of this book is that if students can see the real-life implications of science content, they will be motivated to carry out hands-on, minds-on science investigations and personally care about the results. Science educators have, for decades, emphasized the importance of science experiences for students that emphasize personal involvement in the learning process. I firmly believe that the use of open-ended stories that challenge students to engage in real experimentation about real science content can be a step toward this goal. Furthermore, I believe that students who see a purpose to their learning and experimentation are more likely to understand the concepts they are studying. I sincerely hope that the contents of this book will relieve the overburdened teacher from the exhausting work of designing inquiry lessons from scratch.

These stories feature children in natural situations at home, on the playground, at parties, in school, or in the outdoors. Students should identify with the story characters, to share their frustrations, concerns, and questions. The most important role for the adult is to help guide and facilitate investigations and to debrief activities with them and to think about their analyses of results and conclusions. The children often need help to go to the next level and to develop new questions and find ways of following these questions to a conclusion. Our philosophy of science education is based on the belief that children can and want to care enough about problems to make them their own. This should enhance and invigorate any curriculum. In short, students can begin to lead the curriculum and because of their personal interest in the questions that evolve from their activities, they will maintain interest for much longer than they would if they were following someone else's lead.

A teacher told me that one of her biggest problems is to get her students to "care" about the topics they are studying. She says they go through the motions but without affect. Perhaps this same problem is familiar to you. I hope that this book can help you to take a step toward solving that problem. It is difficult if not impossible to make each lesson personally relevant to every student. However, by focusing on everyday situations and highlighting kids looking at everyday phenomena, I believe that we can come closer to reaching student interests.

I strongly suggest the use of complementary books as you go about planning for inquiry teaching. Five special books are *Uncovering Student Ideas* (volumes 1, 2, 3, and 4) by Page Keeley et al., published by the NSTA Press and *Science Curriculum Topic Study* by Page Keeley, published by Corwin Press and NSTA. The multivolume *Uncovering Student Ideas* helps you to find out what kinds of preconceptions your students bring to your class. *Science Curriculum Topic Study* focuses on finding the background necessary to plan a successful standards-based unit. I would also strongly recommend that you find a copy of *Science Matters: Achieving Scientific Literacy*, by Robert Hazen and James Trefil. This book will become your reference for many scientific matters. It is written in a simple, direct and accurate manner and will give you the necessary background in the sciences when you need it. Finally please acquaint yourself with *Making Sense of Secondary Science: Research Into Children's Ideas* (Driver et al. 1994). The title of this book can be misleading to American teachers, because in Great Britain, anything above primary level is referred to as secondary. It is a compilation of the research done on children's thinking about science and is a must-have for teachers. Use it as a reference in looking for the preconceptions your students probably bring to your classroom.

In 1978, David Ausubel made one of the most simple but telling comments about teaching: "The most important single factor influencing learning is what the learner already knows; ascertain this, and teach him accordingly." The background material that accompanies each story is designed to help you to find out what your learners already know about your chosen topic and what to do with that knowledge as you plan. The above-mentioned books will supplement the materials in this book and deepen your understanding of teaching for inquiry.

How then, is this book set up to help you to plan and teach inquiry-based science lessons?

HOW THIS BOOK IS ORGANIZED

There is a concept matrix (p. 41) that can be used to select a story most related to your content need. Following this matrix you will find the stories and the background material in separate chapters. Each chapter, starting with Chapter 5, will have the same organizational format. First you will find the story, followed by background material for using the story. The background material will contain the following sections:

Purpose

This section describes the concepts and/or general topic that the story attempts to address. In short, it tells you where this story fits into the general scheme of science concepts. It may also place the concepts within a conceptual scheme of a larger idea.

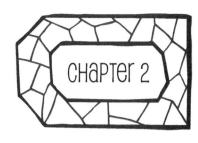

Related Concepts

A concept is a word or combination of words that form a mental construct of an idea. Examples are *condensation, erotion, heat transfer,* and *probability.* Each story is designed to address a single concept but often the stories open the door to several concepts. You will find a list of possible related concepts in the teacher background material. You should also check the matrix of stories and related concepts.

Don't Be Surprised

In most cases, this section will include projections of what your students will most likely do and how they may respond to the story. The projections relate to the content but focus more on the development of their current understanding of the concept. The explanation will be related to the content but will focus more on the development of the understanding of the concept. There will be references made to the current alternative conceptions your students might be expected to bring to class. It may even challenge you to prepare for teaching by doing some of the projected activities yourself, so that you are prepared for what your students will bring to class.

Content Background

This material will be a very succinct "short course" on the conceptual material that the story targets. It will not, of course, be a complete coverage but should give you enough information to feel comfortable in using the story and planning and carrying out the lessons. In most instances, references to books, articles, and internet connections will also help you in preparing yourself to teach the topic. It is important that you have a reasonable knowledge of the topic in order for you to lead the students through their inquiry. It is not necessary, however, for you to be an expert on the topic. Learning along with your students can help you to understand how their learning takes place and make you a member of the class team striving for understanding of natural phenomena.

Table 2.1.
Thematic Crossover Between Stories in This Book and *Uncovering Student Ideas in Science*, Volumes 1–4

Story in this book	*Uncovering Student Ideas in Science*			
	Volume 1	**Volume 2**	**Volume 3**	**Volume 4**
Moon Tricks	Going Through a Phase	Objects in the Sky	n/a	Moonlight; Lunar Eclipse

(continued to next page)

(continued from previous page)

Story in this book	Uncovering Student Ideas in Science			
	Volume 1	**Volume 2**	**Volume 3**	**Volume 4**
What's the Moon Like Around the World?	Gazing at the Moon; Going Through a Phase	Objects in the Sky; Emmy's Moon and Stars	What Is a Hypothesis?	Moonlight
Daylight Saving Time	n/a	Darkness at Night; Objects in the Sky	Is It a Theory? Me and My Shadow; Where Do Stars Go?	n/a
Sunrise, Sunset	n/a	Darkness at Night; Objects in the Sky	Is It a Theory? Me and My Shadow; Where Do Stars Go? Summer Talk	Camping Trip
Now Just Wait a Minute	n/a	n/a	Is It Scientific Inquiry?	n/a
What's Hiding in the Woodpile?	Mitten Problem; Objects and Temperature	n/a	Thermometer	Global Warming
Cool It, Dude!	Cookie Crumbling? Is It Melting? Ice Cubes in a Bag	Ice Cold Lemonade; Freezing Ice	Ice Cubes in a Glass; Thermometer	Ice Water
The New Greenhouse	Objects and Temperature	n/a	Thermometer	Global Warming
Where Did the Puddles Go?	Wet Jeans	n/a	What Is a Hypothesis? What Are Clouds Made Of?	Warming Water
The Little Tent That Cried	Wet Jeans	n/a	Where Did the Water Come From?	Camping Trip

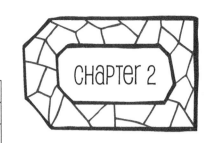

Story in this book	Uncovering Student Ideas in Science			
	Volume 1	Volume 2	Volume 3	Volume 4
Where Are the Acorns?	n/a	n/a	Me and My Shadow; Summer Talk	n/a
The Coldest Time of the Day	n/a	n/a	What Is a Hypothesis?	Camping Trip
Frosty Morning	n/a	Ice Cold Lemonade	n/a	Camping Trip
Master Gardener	Beach Sand; Mountain Age	Is It a Rock #1	n/a	Is It Food?
A Day on Bare Mountain	Talking About Gravity; Beach Sand; Mountain Age	Is It a Rock (1 & 2); Mountaintop Fossil	Earth's Mass	n/a
What Are the Chances?	n/a	n/a	What Is a Hypothesis?	Where Would It Fall?
Here's the Crusher	n/a	n/a	What Is a Hypothesis?	n/a
Rotten Apples	n/a	Is It Food for Plants?	Earth's Mass; Rotting Apple	Is It Food? Is It a System?
Is the Earth Getting Heavier?	n/a	n/a	Earth's Mass; Rotting Apple; What Is a Hypothesis?	Is It a System?

Related Ideas from the National Science Education Standards *(NRC 1996) and* Benchmarks for Science Literacy *(AAAS 1993)*

These two documents are considered to be the National Standards upon which most of the local and state standards documents are based. For this reason, the concepts listed for the stories are almost certainly the ones listed to be taught in your local curriculum. It is possible that some of the concepts are not mentioned specifically in the Standards but are clearly related. I suggest that you obtain a copy of *Curriculum Topic Study* (Keeley 2005), which will help you immensely with finding information about content, children's preconceptions, standards, and more resources. Even though it may not be mentioned specifically in each of the stories, you can assume that all of the stories will have connections to the Standards and Benchmarks in the area of Inquiry, Standard A.

Using the Story With Grades K–4 and 5–8

These stories have been tried with children of all ages. We have found that the concepts apply to all grade levels but at different levels of sophistication. Some of the characters in the stories have themes and characters that resonate better with one age group than another. However, the stories can be easily altered to appeal to an older or younger group by changing the characters to a more appropriate age or using slightly different age-appropriate dialog. The theme should be the same; just the characters and setting modified. Please read the suggestions for both grade levels.

As you may remember from the case study in the introduction, grade level is of little consequence in determining which stories are appropriate at which grade level. Both classes developed hypotheses and experiments appropriate to their developmental abilities. Second graders were satisfied to find out what happens to the length of a tree's shadow over a school year while the fifth-grade class developed more sophisticated experiments involving length of day, direction of shadows over time and the daily length of shadows over an entire year. The main point here is that by necessity some stories are written with characters more appealing to certain age groups than others. Once again, I encourage you to read both the K–4 and 5–8 sections of Using the Story, because ideas presented for either grade level may be suited to your particular students.

There is no highly technical apparatus required. Readily available materials found in the kitchen, bathroom, or garage will usually suffice. Each chapter includes background information about the principles and concepts involved and a list of materials you might want to have available. These suggestions of ideas and materials are based upon our experience while testing these stories with children. While we know that classrooms, schools, and children differ, we feel that most childhood experiences and development result in similar reactions to explaining and developing questions about the tales. The problems beg for solutions and most importantly, create new questions to be explored by your young scientists.

Here you will find suggestions to help you teach the lessons that will allow your students to become active inquirers develop their hypotheses, and finally finish the story that you may remember was left open for just this purpose. I have not listed a step-by-step approach or set of lesson plans to accomplish this end. Obviously, you know your students, their abilities, their developmental levels, their learning abilities and disabilities better than anyone. You will find however, some suggestions and some techniques that we have found work well in teaching for inquiry. You may use them as written or modify them to fit your particular situation. The main point is that you try to involve your students as deeply as possible in trying to solve the mysteries posed by the stories.

Related Books and NSTA Journal Articles

Here, we will list specific books and articles from the constantly growing treasure trove of National Science Teacher Association (NSTA) resources for teachers. While our listings are not completely inclusive, you may access the entire scope of resources on the internet at *www.nsta.org/store*. Membership in NSTA will allow you to read all articles online free of charge.

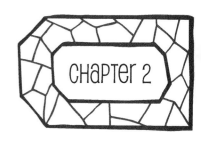

References

References will be provided for the articles and research findings cited in the background section for each story.

Concept Matrix

At the beginning of the story section you will find a concept matrix listing the concepts most related to each story. It can be used to select a story that matches your instructional needs.

FINAL WORDS

I was pleased to find that Michael Padilla, past president of NSTA, asked the same questions as I did when I decided to write a book that focused on inquiry. In the May 2006 edition of *NSTA Reports*, Mr. Padilla in his "President's Message" commented, "To be competitive in the future, students must be able to think creatively, solve problems, reason and learn new, complex ideas… [Inquiry] is the ability to think like a scientist, to identify critical questions to study; to carry out complicated procedures, to eliminate all possibilities except the one under study; to discuss, share and argue with colleagues; and to adjust what you know based on that social interaction." Further, he asks, "Who asks the question?…Who designs the procedures?…Who decides which data to collect?…Who formulates explanations based upon the data?…Who communicates and justifies the results?…What kind of classroom climate allows students to wrestle with the difficult questions posed during a good inquiry?"

I believe that this book speaks to these questions and that the techniques proposed here are one way to answer the above questions with, "The students do!" in the kind of science classroom this book envisions.

REFERENCES

Ausubel, D., J. Novak, and H. Hanensian. 1978. *Educational psychology: A cognitive view.* New York: Holt, Rinehart, and Winston.

Driver, R., A. Squires, P. Rushworth, and V. Wood-Robinson. 1994. *Making sense of secondary science: Research into children's ideas.* London and New York: Routledge Falmer.

Hazen, R., and J. Trefil. 1991. *Science matters: Achieving scientific literacy.* New York: Anchor Books.

Keeley, P. 2005. *Science curriculum topic study: Bridging the gap between standards and practice.* Thousand Oaks, CA: Corwin Press.

Keeley, P., F. Eberle, and C. Dorsey. 2008. *Uncovering student ideas in science, volume 3: Another 25 formative assessment probes.* Arlington, VA: NSTA Press.

Keeley, P., F. Eberle, and L. Farrin. 2005. *Uncovering student ideas in science, volume 1: 25 formative assessment probes.* Arlington, VA: NSTA Press.

Keeley, P., F. Eberle, and J. Tugel. 2007. *Uncovering student ideas in science, volume 2: 25 more formative assessment probes.* Arlington, VA: NSTA Press.

Keeley, P., and J. Tugel. 2009. *Uncovering student ideas in science, volume 4: 25 new formative assessment probes.* Arlington, VA: NSTA Press.

Konicek-Moran, R. 2008. *Everyday science mysteries: Stories for inquiry-based science teaching.* Arlington, VA: NSTA Press.

Konicek-Moran, R. 2009. *More everyday science mysteries: Stories for inquiry-based science teaching.* Arlington, VA: NSTA Press.

Padilla, M. 2006. President's message. *NSTA Reports* 18 (9): 3.

CHAPTER 3

USING THIS BOOK IN DIFFERENT WAYS

althougn the book was originally designed for use with K–8 students by teachers or adults in informal settings, it became obvious that a book containing stories and content material for teachers intent on teaching in an inquiry mode had other potential uses. I list a few of them below to show that the book has several uses beyond the typical elementary and middle school population in formal settings.

USING THE BOOK AS A CONTENT CURRICULUM GUIDE

When asked by the University of Massachusetts to teach a content course for a special master's degree program in teacher education, I decided to use *Everyday Science Mysteries* as one of several texts to teach content material. A major premise in the book is that students, when engaged in answering their own questions will delve into a topic at a level commensurate with their intellectual development and learning skills. Therefore, even though the stories were designed for people younger than themselves, the students in the class were able to find questions to answer that were at a level of sophistication that challenged them.

During the fall 2007 semester this book was used as a text and curriculum guide for a class titled Exploring the Natural Sciences Through Inquiry at the University of Massachusetts in Amherst. The shortened version of the syllabus for the course follows:

Exploring the Natural Sciences Through Inquiry
EDUC 692 O
Fall 2007

Instructor: Dr. Richard D. Konicek, Professor Emeritus

Course Description:
This course is designed for elementary and middle school teachers who need, not only to deepen their content knowledge in the natural sciences, but also to understand how inquiry can be used in the elementary and middle school classroom. Natural sciences mean the Biological Sciences, Earth and Space Sciences, and the Physical Sciences. Teachers will sample various topics from each of the above areas of science through inquiry techniques. The topics will be chosen from everyday phenomena such as Astronomy (Moon and Sun observations), Physics (motion, energy, thermodynamics, sound periodic motion), and Biology (botany, zoology, animal and plant behavior, evolution).

Course Objectives:
It is expected that each student will:
- Gain content background in each of the three areas of natural science.
- Be able to apply this content to their teaching methods.
- Develop questions concerning a particular phenomenon in nature.
- Design and carry out experiments to answer their questions.
- Analyze experimental data and draw conclusions.
- Consult various sources to verify the nature of their conclusions.
- Read scientific literature appropriate to their studies.
- Extend their knowledge to use with middle school children both in content and methodology.

Relationship to the Conceptual Framework of the School of Education:

Collaboration:	Teachers will work in collaborative teams during class meetings to acquire science content and pedagogical knowledge and skills.
Reflective Practice:	Teachers will develop and implement formative assessment probes with their students.
Multiple Ways of Knowing:	Teachers will share science questions and their methods of inquiry chosen to answer those questions.
Access, Equity, and Fairness:	Teachers reflect on student understandings based on students' stories.
Evidence-Based Practice:	Teachers will explore formative assessment through the use of probes.

Required Texts:
Hazen, R. M., and J. Trefil. 1991. *Science matters*. New York: Anchor Books.
Keeley, P., F. Eberle, and J. Tugel. 2007. *Uncovering student ideas in science: 25 more formative assessment probes, vol. 2.* Arlington, VA: NSTA Press. Konicek-

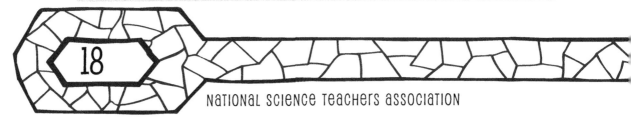

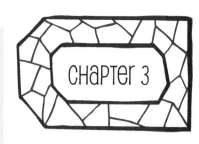

Moran, R. 2008. *Everyday science mysteries.* Arlington, VA: NSTA Press.

Resource Texts:

American Association for the Advancement of Science (AAAS). 2001. *Atlas of science literacy* (vol. 1). Washington, DC: Project 2061.

American Association for the Advancement of Science (AAAS). 2007. *Atlas of science literacy* (vol. 2). Washington, DC: Project 2061.

Driver, R., A. Squires, P. Rushworth, and V. Wood-Robinson. 1994. *Making sense of secondary science.* London: Routledge-Falmer.

Keeley, P., F. Eberle, and L. Farrin. 2005. *Uncovering student ideas in science, vol. 1.* Arlington, VA: NSTA Press.

Topics To Be Investigated in Volume One:

Everyday Science Mysteries is organized around stories. The core concepts related to the National Science Education Standards developed by the National Research Council in 1996 are the basis for the concept selection. The story titles and related core concepts are shown in the matrices below.

Earth Systems Science

Core Concepts	Stories				
	Moon Tricks	Where Are the Acorns?	Master Gardener	Frosty Morning	The Little Tent That Cried
States of Matter			X	X	X
Change of State			X	X	X
Physical Change			X	X	X
Melting			X	X	
Systems	X	X	X	X	X
Light	X	X			
Reflection	X	X		X	
Heat Energy			X	X	X
Temperature				X	X
Energy			X	X	X
Water Cycle				X	X
Rock Cycle			X		
Evaporation				X	X
Condensation				X	X
Weathering			X		
Erosion			X		
Deposition			X		
Rotation/Revolution	X	X			
Moon Phases	X				
Time	X	X			

Physical Sciences

Core Concepts	Magic Balloon	Bocce Anyone?	Grandfather's Clock	Neighbor-hood Telephone Service	How Cold Is Cold?
Energy	X	X	X	X	X
Energy Transfer	X	X	X	X	X
Conservation of Energy		X			X
Forces	X	X	X		
Gravity	X	X	X		
Heat	X				X
Kinetic Energy		X	X		
Potential Energy		X	X		
Position and Motion		X	X		
Sound				X	
Periodic Motion			X	X	
Waves				X	
Temperature	X				X
Gas Laws	X				
Buoyancy	X				
Friction		X	X		
Experimental Design	X	X	X	X	X
Work		X	X		
Change of State					X
Time		X	X		

Biological Sciences

Core Concepts	About Me	Bugs	Dried Apples	Seed Bargains	Trees From Helicopters
Animals	X	X			
Classification		X	X	X	X
Life Processes	X	X	X	X	X
Living Things	X	X	X	X	X
Structure and Function		X	X		X
Plants			X	X	X

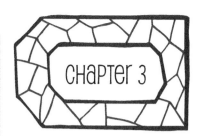

Adaptation		X			X
Genetics/ Inheritance	X		X	X	X
Variation	X		X	X	X
Evaporation			X		
Energy		X	X	X	X
Systems	X	X	X		X
Cycles	X	X	X	X	X
Reproduction	X	X	X	X	X
Inheritance	X	X	X		X
Change		X	X		
Genes	X		X		X
Metamorphosis		X			
Life Cycles		X	X		X
Continuity of Life	X	X	X	X	X

Assignments:

Astronomy (25%): Everyone will be expected to explore the daytime astronomy sequence, which will aim to develop models of the Earth, Moon, and Sun relationships. Students will keep a Moon journal and Sun shadow journal over the course of the semester, which they will turn in periodically.

Topics (50%): In addition, students will pick at least two topics from each of the Earth, Physical and Biological areas for study during the semester. Students will come up with a topic question and do an investigation or experiment regarding the topic questions posed. (For example: Are there acorns that do not need a dormancy period before germinating?) These questions and experiments will be shared with the class as they progress so that all students will either be directly involved in learning about the content or indirectly involved by listening to reports and critiquing those reports. In addition to the experiments, students will (1) involve their students in their experiments/investigations and (2) design and give formative assessment probes to their students to find out what knowledge they already possess. Students will be graded on their experimental designs, their presentations of their data and upon their conclusions. I will develop a rubric with the students that will address the goals stated above and their values to be calculated for their grades.

Attendance/Participation (25%): Attendance at all course meetings is required.

References for Course Development:

American Association for the Advancement of Science (AAAS).1993. *Benchmarks for science literacy.* New York: Oxford University Press.

Ausubel, D., J. Novak, and H. Hanensian. 1978. *Educational psychology: A cognitive view.* New York: Holt, Rinehart and Winston.

Bransford, J. D., A. L. Brown, and R. R. Cocking, eds. 1999. *How people learn.* Washington, DC: National Academy Press.

Duckworth, E. 1986. *Inventing density.* Grand Forks, ND: Center for Teaching and Learning, University of North Dakota.

Driver, R., A. Squires, P. Rushworth, and V. Wood-Robinson. 1994. *Making sense of secondary science: Research into children's ideas.* London and New York: Routledge Falmer.

Hazen, R., and J. Trefil. 1991. *Science matters: Achieving scientific literacy.* New York: Anchor Books.

Keeley, P. 2005. *Science curriculum topic study: Bridging the gap between standards and practice.* Thousand Oaks, CA: Corwin Press.

Keeley, P., F. Eberle, and L. Farrin. 2005. *Uncovering student ideas in science: 25 formative assessment probes, volume 1.* Arlington, VA: NSTA Press.

Keeley, P., F. Eberle, and J. Tugel. 2007. *Uncovering student ideas in science: 25 more formative assessment probes, volume 2.* Arlington, VA: NSTA Press.

Konicek-Moran, R. 2008. *Everyday science mysteries.* Arlington, VA: NSTA Press.

Martin, K., and E. Miller. 1990. Storytelling and science. In *Toward a whole language classroom: Articles from language arts,* ed. B. Kiefer, 1986–1989. Urbana, IL: National Council of Teachers of English.

National Research Council (NRC). 2000. *Inquiry and the national science education standards: A guide for teaching and learning.* Washington, DC: National Academies Press.

Osborne, R., and P. Fryberg. 1985. *Learning in science: The implications of children's science.* Auckland, New Zealand: Heinemann.

Schneps, M. 1996. *A private universe project.* Washington, DC: Harvard Smithsonian Center for Astrophysics.

Shapiro, B. 1994. *What children bring to light.* New York: Teachers College Press.

Watson, B., and R. Konicek. 1990. Teaching for conceptual change: Confronting children's experience. *Phi Delta Kappan* May: 680–684.

The course was taught as a graduate course for teachers or prospective teachers of elementary or middle school students. The course could be classified as a content/pedagogy class for teachers who had minimal science backgrounds as well as minimal skills in teaching for inquiry. My premise was that if teachers would learn content through inquiry techniques, they would be convinced of their efficacy as learning techniques and would be likely to use them to teach content, in their own classes. As it turned out, those teacher-students who had classes of their own and were full-time teachers did work on their projects with their students with very satisfactory results according to the teachers. As a result, both teachers and students were learning science content through inquiry techniques. Because the teachers in the class were completing an assignment, they were able to be honest with their students about not knowing the outcome of their investigations. This is

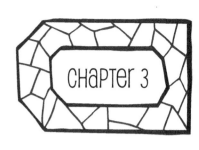

often a problem with teachers who are afraid to admit that they are learning along with the students. In this case, the students were excited about learning along with their teachers and vice versa. Teachers with classrooms were also able to develop rubrics with their students for the grading of their explorations and therefore were involved with some metacognition as well.

As a result of this small foray into the use of the book in this manner, I am convinced that the book can be used as a content guide for undergraduate and graduate content-oriented courses for teachers. As noted in the syllabus, the use of other supplementary texts for content and pedagogy add to the strength of the course in preparing teachers to use inquiry techniques and to learn content themselves. With the use of the internet, very little information is hidden from anyone with minimum computer skills. Unlike many survey courses chosen by teachers who are science-phobic, this course did not attempt to cover a great number of topics but to teach a few topics for understanding. The basic premise is that when deciding between coverage and understanding science topics and concepts, understanding wins every time. It is well known that our current curriculum in the United States has been faulted for being a mile wide and an inch deep. High-stakes testing seems to also add to the problem since almost all teachers whom I have interviewed over the last few years are reluctant to teach for understanding using inquiry methods because teaching for understanding takes more time and does not allow for coverage of the almost infinite amount of material that might appear on standardized tests. Thus, student misconceptions are seldom addressed and continue to persist even though students can do reasonably well on teacher-made tests and assessment tools and still hold onto their misconceptions. See Bonnie Shapiro's book, *What Children Bring to Light* (1994).

USING THIS BOOK as a resource BOOK FOR SCIENCE METHODS COURSES IN Teaching Preparation Programs

Traditionally, science methods courses in the United States are taught to classes mainly composed of science-phobic students. One of the main goals of science methods courses is to make students comfortable with science teaching and to help students develop skills in teaching science to youngsters using a hands-on, minds-on approach. Unfortunately, a great many students come to these methods courses with a minimum of science content courses and many of those are either survey (non-laboratory) courses or courses taught in a large lecture format. In 12–13 weeks, methods instructors are expected to convert these students into confident, motivated teachers who are familiar with techniques that promote inquiry learning among their students. Having taught this type of course to undergraduates and career-changing graduate students for over 30 years, I have found that making

students comfortable with science is the first goal. This is often accomplished by assigning students science tasks that can be accomplished with a minimum of stress and with a maximum of success. Second, I try to instill the ideas commensurate with the nature of science as a discipline. Third, I find that it is often necessary to teach a little content for those who are rusty and need to clarify some of their own misconceptions. Lastly, but not least important, I try to acquaint them with resources in the field so that they know what is available to them as they enter their teaching careers. Obviously, here is an opportunity to acquaint them with current information about the learners themselves, how they learn, and how to teach for inquiry.

As a final assignment for my methods classes, I assign the students the task of writing an everyday science mystery and a paper to accompany it, which will describe how they will use the story to teach a concept using the inquiry approach. The results have far exceeded what I had been receiving from the typical lesson plan used by others and me through the years. This book would not only provide the text on teaching science found in the early chapters but would provide a model for producing everyday science mysteries for topics of the students' choices.

USING THE STORIES AS INTERACTIVE INQUIRY PLAYS

Due to the innovation of the teachers of Knox County, Tennessee, and the actions of instructional coaches Andrea Allen and Theresa Nixon, a new and exciting method of introducing the stories has been invented. These teachers have adapted the mystery stories into a theatrical mode called the "Everyday Science Mystery Readers Theater." They invite teachers to make an interactive play out of the mystery stories instead of reading them. This involves the students in acting out the stories and in doing so, puts them further into the mysteries. We thank them for this innovation and invite you all to try this with your students. See Chapter 4: Science and Literacy for more information on student reading and writing in science. One of the plays, "Rotten Apples," is reproduced here with the teachers' kind permission (see Chapter 22 for the original story and discussion).

SCIENCE MYSTERY THEATER PRESENTS: ROTTEN APPLES

Characters:
Ted
Steve
Narrator

Setting: Apple Orchard

Narrator: Ted and Steve were on their way home from school one October day and decided to take a shortcut, as they often did, through the old apple orchard near their homes.

Ted: You know Steve all of our homes are built on land that used to be part of a huge apple orchard. The owner sold off a lot of his old orchard to developers who built these homes. My mom remembers when all of this land, including our school, was an apple orchard.

Steve: Yeah, I remember hearing about that, too.

Ted: I wonder what happened to all of the old apples that were all over this field and around our homes since the trees were cut down.

Steve: And look around this orchard, now. There are all kinds of apples lying on the ground that didn't get picked or just fell after the harvest.

Ted: You'd think that the apples from all of the years that trees have been dropping them would be around here and we'd be up to our knees in old apples. But actually I wonder where they all go to. Does someone come out here and clean them all up?

Steve: I don't think so, unless they do that to make apple cider. Yeah that's probably what happens, or else in the spring we'd be walking over a lot of apples.

Narrator: It was a nice warm fall day and the boys took their time walking home and stopped to look at some of the apples left on the ground.

Ted: Man, look at these apples. They look half-rotten already. They wouldn't make cider out of these would they? There're all goopy and look like they have worms or something in them. They're not good for anything anymore.

Steve: I bet animals eat a lot of them but not all of them, 'cause there are still a lot of them left. So why aren't they here in the spring? And where are the apples around our school and our houses?

Ted: I hear that they turn into soil.

Steve: Just like that? Magic? How can that be, soil is soil and it's always here, with or without apples. Soil is dirt, right, so apples can't turn into dirt! It has to be more complicated than that.

Ted: I know! Let's take some of them home and put them in my yard to see what happens. We'll put them outside where the dogs can't get at them and we'll keep an eye on them. Nothing like good ole observation, like Ms. Green keeps telling us in science class.

Steve: Okay as long as it's in your yard. I don't think my folks will like rotten apples in my yard and our dog will eat them for sure. She eats anything, and I mean *anything*!

Narrator: And so the boys took a bunch of apples from the ground and took them home for some good old observation, just like Ms. Green kept telling them. And next spring …?

Use for Homeschool Programs

Homeschooling parents have a great many resources at their disposal, as any internet search will show. Curricular suggestions and materials are available for those parents and children who choose to conduct their education at home. Science is one of those subjects that might be difficult for many parents whose science backgrounds are a bit weak or outdated. Parents and children working together to solve a story-driven mystery could use this book easily. The connections to the national Standards and the Benchmarks in science also help in making sure that the home schooling curriculum is uncovering the nationally approved scientific concepts. Parents would use the book just as any teacher would use it except there would be fewer opportunities for class discussions and the parents would have to do a bit more discussion with their children to solidify their understanding of their investigations.

Reference

Shapiro, B. 1994. *What children bring to light*. New York: Teachers College Press.

CHAPTER 4

SCIENCE AND LITERACY

While heading into the final chapter before launching into the stories, I couldn't resist introducing you to a piece of literature that is seldom read except by English majors. The quotation that follows is from Irish novelist James Joyce in his classic book *Ulysses*, written in 1922:

> Where was the chap I saw in that picture somewhere? Ah yes, in the dead sea, floating on his back, reading a book with a parasol open. Couldn't sink if you tried: so thick with salt. Because the weight of the water, no, the weight of the body in the water is equal to the weight of the what? Or is it the volume is equal to the weight? It's a law something like that. Vance in High school cracking his fingerjoints, teaching. The college curriculum. Cracking curriculum. What is weight really when you say the weight? Thirtytwo feet per second, per second. Law of falling bodies: per second, per second. They all fall to the ground. The earth. It's the force of gravity of the earth is the weight. (p. 73)

In the novel, Joyce's main character Bloom recalls a picture of someone floating in the Dead Sea, and tries to recall the science behind it. Have you or have you observed others who, while trying to recall something scientific, resorted to a mishmash of scientic knowledge,

half-remembered and garbled? (For this foray into literature, I am indebted to Michael J. Reiss, who called my attention to this passage in an article of his in *School Science Review*.)

In his school days, Bloom seems to have been fascinated both with the curriculum and the teacher in his physics class. However, Bloom's memory of the science behind buoyancy runs the gamut from unrelated science language pouring out of his memory bank to visions of his teacher cracking his finger joints. Unfortunately, even today, this might well be the norm rather than the exception. This phenomenon is exactly what we are trying to avoid in our modern pedagogy and now leads us to the main point of this chapter.

There are many ways of connecting literacy and science. We shall look briefly at the research literature and find some ideas that will make the combination of literacy and science not only worthwhile but also essential for learning.

LITERACY AND SCIENCE

In pedagogical terms there are differences between scientific literacy and the curricular combination of science and literacy, but perhaps they have more in common than one might expect. *Scientific literacy* is the ability to understand scientific concepts so that they have a personal meaning in everyday life. In other words, a scientifically literate population can use their knowledge of scientific principles in situations other than those in which they learned them. For example, I would consider people scientifically literate if they were able to use their understanding of ecosystems and ecology to make informed decisions about saving wetlands in their community. This is of course, what we would hope for in every aspect of our educational goals regardless of the subject matter. *Literacy* refers to the ability to read, write, speak, and make sense of text. Since most schools emphasize reading, writing, and mathematics, they often take priority over all other subjects in the school curriculum. How often have I heard teachers say that their major responsibility is reading and math, and that there is no time for science? But there is no need for competition for the school day. I believe that this misconception is caused by the lack of understanding of the synergy created by integration of subjects. In *synergy,* you get a combination of skills that surpasses the sum of the individual parts.

So what does all of this have to do with teaching science as inquiry? There is currently a strong effort to combine science and literacy. One reason is that there is a growing body of research that stresses the importance of language in learning science. "Hands-on" science is nothing without its "minds-on" counterpart. I am fond of reminding audiences that a food fight is a hands-on activity, but one does not learn much through mere participation, except perhaps the finer points of the aerodynamic properties of Jell-O. The understanding of scientific principles is not imbedded in the materials themselves or in the manipulation of these materials. Discussion, argumentation, discourse of all kinds, group consensus and social interaction—all forms of communication are necessary for students to make meaning out of the activities in which they have engaged. And these require *language* in the form of writing, reading, and particularly speaking. They require that students think about their thinking—that they hear their own and others' thoughts and ideas spoken out loud and perhaps eventually see them in writing to make sense of what they have been doing and the results they have been getting in their activities. This is the often forgotten "minds-on" part of the "hands-on, minds-on" couplet. Consider the following:

> In schools, talk is sometimes valued and sometimes avoided, but—and this is surprising—talk is rarely taught. It is rare to hear teachers discuss their efforts to teach students to talk well. Yet talk, like reading and writing, is a major motor—I could even say the major motor—of intellectual development. (Calkins 2000, p. 226)

For a detailed and very useful discussion of talk in the science classroom, I refer you to Jeffrey Winokur and Karen Worth's chapter, "Talk in the Science Classroom: Looking at What Students and Teachers Need to Know and Be Able to Do" in *Linking Science and Literacy in the K–8 Classroom* (2006). Also check out Chapter 8 in this book. There is also recent evidence that ELL learners gain a great deal from talking, in both their science learning and new language acquisition (Rosebery and Warren 2008).

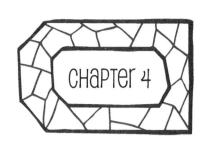

Linking inquiry-based science and literacy has strong research support. First, the conceptual and theoretical work of Padilla and his colleagues suggest that inquiry science and reading share a set of intellectual processes (e.g., observing, classifying, inferring, predicting, and communicating) and that these processes are used whether the student is conducting scientific experiments or reading text (Padilla, Muth, and Padilla 1991). Helping children become aware of their thinking as they read and investigate with materials will help them understand and practice more *metacognition*.

You, the teacher, may have to model this for them by thinking out loud yourself as you view a phenomenon. Help them to understand why you spoke as you did and why it is important to think about your process of thinking. You may say something like, "I think that warm weather affects how fast seeds germinate. I think that I should design an experiment to see if I am right." Then later, "Did you notice how I made a prediction that I could test in an experiment?" Modeling your thinking can help your students see how and why the talk of science is used in certain situations.

Science is about words and their meanings. Postman made a very interesting statement about words and science. He said "Biology is not plants and animals. It is language about plants and animals…. Astronomy is not planets and stars. It is a way of talking about planets and stars" (1979, p. 165). To emphasize this point even further, I might add that science is a language, a language that specializes in talking about the world and being in that world we call science. It has a special vocabulary and organization. Scientists use this vocabulary and organization when they talk about their work. Often, it is called "discourse" (Gee 2004). Children need to learn this discourse when they present their evidence, when they argue the fine points of their work, evaluate their own and others' work and refine their ideas for further study.

Students do not come to you with this language in full bloom; in fact the seeds may not even have germinated. They attain it by doing science and being helped by knowledgeable adults who teach them about controlling variables, conducting fair tests, having evidence to back up their statements, and using the processes of science in their attempts at what has been called "firsthand inquiry" (Palincsar and Magnusson 2001). This is inquiry that uses direct involvement with materials, or in other more familiar words, the hands-on part of scientific investigation. The term *secondhand investigations* refers to the use of textual matter, lectures, reading data, charts, graphs, or other types of instruction that do not feature direct contact with materials. Cervetti et al. (2006) put it so well:

> [W]e view firsthand investigations as the glue that binds together all of the linguistic activity around inquiry. The mantra we have developed for ourselves in helping students acquire conceptual knowledge and the discourse in which that knowledge is expressed (including particular vocabulary) is "read it, write it, talk it, do it!"—and in no particular order, or better yet, in every possible order. (p. 238)

So you can see that it is also important that the students talk about their work; write about their work; read about what others have to say about the work they are doing, in books or via visual media; and take all possible opportunities to document their work in a way that is useful to them in looking back at what they have found out about their work.

THE LANGUAGE OF SCIENCE

Of course, writing, talking, and reading in the discipline of science is different than other disciplines. For example, science writing is simple and focuses on the evidence obtained to form a conclusion. But science includes things other than just verbal language. It includes tactile, graphic, and visual means of designing studies, carrying them out, and communicating the results to others. Also important is that science has many unfamiliar words; many common words such as *work, force, plant food, compound,* and *density* have different meanings in the real world of the student but have precise and often counterintuitive meanings in science. For example, if you push against a car for 30 minutes until perspiration runs off your face, you feel as though you have "worked" hard even though the car has not budged a centimeter. In physics, unless the car has moved, you have done no work at all. We tell students that plants make their own food and then show them a bag of "plant food." We tell children to "put on warm clothes," yet the clothes have nothing to do with producing warmth.

Students have to change their way of communicating when they study science. They must learn new terminology and clarify old terms in scientific ways. We as teachers can help in this process by realizing that we are not just science teachers but also language teachers. When we talk of scientific things we talk about them in the way the discipline works. We should not avoid scientific terminology but try to connect it whenever possible to common metaphors and language. We should use pictures and stories.

We need also to know that science contains many words that ask for thought and action on the part or the students. Sentences with words like *compare, evaluate, infer, observe, modify,* and *hypothesize* prompt students to solve problems. We can only teach good science by realizing that language and intellectual development go hand in hand and that one without the other is mostly meaningless.

SCIENCE NOTEBOOKS

Many science educators have lately touted science notebooks as an aid to students involving themselves more in the discourse of science (Campbell and Fulton 2003). Their use has also shown promise in helping English language learners (ELLs) in the development of language skills as well as learning science concepts and the nature of science.

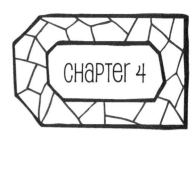

Science notebooks differ from science journals and science logs in that they are not merely for recording data (logs) or reflections of learning (journals), but are meant to be used continuously for recording experimentation, designs, plans, thinking, vocabulary, and concerns or puzzlement. The science notebook is the recording of past, and present thoughts and predictions and are unique to each student. The teacher makes sure that the students have ample time to record events and to also ask for specific responses to such questions as, "What still puzzles you about this activity?"

For specific ideas for using science notebooks and for information on the value of using the notebooks in science, see *Science Notebooks: Writing About Inquiry* by Brian Campbell and Lori Fulton (2003). You can assume that science notebooks are a given in what I envision as an inquiry-oriented classroom. While working in an elementary school years ago, I witnessed some minor miracles of children writing to learn. The most vital lesson for us as teachers was the importance of asking children to write each day about something that still confused them. The results were remarkable. As we read their notebooks, we witnessed their metacognition, and their solutions through their thinking "out loud" in their writing.

The use of science notebooks should be an opportunity for the students to record their mental journey through their activity. Using the stories in this book, the science notebook would include the specific question that the student is concerned with, the lists of ideas and statements generated by the class after the story is read, pictures or graphs of data collected by the student and class, and perhaps the final conclusions reached by the student or class as they try to solve the mystery presented by the story.

Let us imagine that your class has reached a conclusion to the story they have been using and have reached consensus on that conclusion. What options are open to you as a teacher for asking the students to finalize their work? At this juncture, it may be acceptable to have the students actually write the "ending" to the story or write up the conclusions in a standard lab report format. The former method, of course, is another way of actually connecting literacy and science. Many teachers prefer to have their students at least learn to write the "boiler plate" lab reports, just to be familiar with that method, while others are comfortable with having their students write more anecdotal kinds of reports. My experience is that when students write their conclusions in an anecdotal form, while referring to their data to support their conclusions, I am more assured that they have really understood the concepts they have been chasing rather than filling in the blanks in a form. In the end, it is up to you, the classroom teacher, to decide. Of course, it could be done both ways.

As mentioned earlier, a major factor in designing these stories and follow-up activities is based upon one of the major tenets of a philosophy called *constructivism*. That tenet is that knowledge is constructed by individuals in order to make sense of the world in which they live. If we believe this, then the knowledge that each individual brings to any situation or problem must be factored into the way that person tries to solve that problem. By the same token, it is most important to realize that the *identification* of the problem and the way the problem is *viewed* are also factors determined by each individual. Therefore it is vital that the adult facilitator encourage the students to bring into the open, orally and in writing, those ideas they already have about the situation being discussed. In bringing these preconceptions out of hiding, so to speak, all of the children and the teacher can begin playing with all of the cards exposed and alternative ideas about topics can be addressed. Data can be then analyzed openly without any hidden agendas in childrens' minds to sabotage learning. You can find more about this process in the series *Uncovering Student Ideas in Science: 25 Formative Assessment Probes*, vols. 1–4 (2005, 2007, 2008, 2009).

The stories also point out that science is a social, cultural, and therefore human enterprise. The characters in our stories usually enlist others in their investigations, their discussions, and their questions. These people have opinions and hypotheses and are consulted, involved, or drawn into an active dialectic. Group work is encouraged, which in a classroom would suggest cooperative learning. At home, siblings and parents may become involved in the activities and engage in the dialectic as a family group.

The stories can also be read to the children. In this way children can gain more from the literature than if they had to read the stories by themselves. A child's listening vocabulary is usually greater than his or her reading vocabulary. Words that are somewhat unfamiliar to them can be deduced by the context in which they are found. Or, new vocabulary words can be explained as the story is read. We have found that children are always ready to discuss the stories as they are read and therefore become more involved as they take part in the reading. So much the better because getting involved is what this book is all about; getting involved in situations that beg for problem finding, problem solving, and construction of new ideas about science in everyday life.

HELPING YOUR STUDENTS DURING INQUIRY

How much help should you give to your students as they work through the problem? A good rule of thumb is that you can help them as much as you think necessary as long as the children are still finding the situation problematic. In other words, the children should not be following your lead but their own lead. If some of these leads end up in dead ends, then that aspect of scientific investigation is part of their experience too. Science is full of experiences which are not productive. If children read popular accounts of scientific discovery, they could get the impression that the scientist gets up in the morning, says, "What will I discover today?" and then sets off on a clear, straight path to an elegant conclusion before suppertime rolls around. Nothing could be further from the truth! But it is very important to note that a steady diet of frustration can dampen students' enthusiasm for science.

Dead ends can be viewed as signaling a need to develop a new plan or ask the question in a different way. Most important, dead ends should not be looked upon as failures. They are more like opportunities to try again in a different way with a

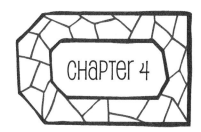

clean slate. The adult's role is to keep a balance so that motivation is maintained and interest continues to flourish. Sometimes this is more easily accomplished when kids work in groups. Most often nowadays, scientists work in teams and use each other's expertise in a group process,

Many people do not understand that the scientific process includes luck, personal idiosyncrasies, and feelings, as well as the so-called *scientific method*. The term *scientific method* itself sounds like a recipe guaranteed to produce success. The most important aid you can provide for your students is to help them maintain their confidence in their ability to do problem solving using all of their ways of knowing. They can use metaphors, visualizations, drawings, or any other method with which they are comfortable to develop new insights into the problem. Then they can set up their study in a way that reflects the scientific paradigm including a simple question, controlling variables, and isolating the one variable they are testing.

Next, you can help them to keep their experimental designs simple and carefully controlled. Third, you can help them to learn keep good data records in their science notebooks. Most students don't readily see the need for this last point, even after they have been told. They don't see the need because the neophyte experimenter has not had much experience with collecting usable data. Until they realize that unreadable data or necessary data not recorded can cause a problem, they see little use for them. The problem is that they don't see it as a problem. Children don't see the need for keeping good shadow length records because they are not always sure what they are going to do with them in a week or a month from now. If they are helped to see the reasons for collecting data and that these data are going to be evidence of a change over time, then they will see the purpose of being able to go back and revisit the past in order to compare it to the present. In this way they can also see the reasons for keeping a log in the first place.

In experiences we have had with children, forcing them to use prescribed data collection worksheets has not helped them to understand the reasons for data collection at all and in some cases has actually caused more confusion or amounted to little more than busy work. On one occasion while circulating around a classroom where children were engaged in a worksheet-directed activity, an observer asked a student what she was doing. The student replied without hesitation, "step 3." Our goal is to empower students engaged in inquiry to the point where they are involved in the activity at a level where all of the steps, including step 3, are designed by the students themselves and for good reason—to answer their own questions in a logical, sequential, meaningful manner. We believe it can be done but it requires patience on the part of the adult facilitators and faith that the children have the skills to carry out such mental gymnastics, with a little help from their friends and mentors.

One last word about data collection. After spending years being a scientist and working with scientists, one common element stands out for me. Scientists keep on their person a notebook that is used numerous times during the day to record interesting items. The researcher may come across some interesting data that may not seem directly connected to the study at the time but he or she makes some notes about it anyway because that entry may come in handy in the future. Memory is viewed as an ephemeral thing, not to be trusted. Scientists' notebooks are a treasured and essential part of the scientific enterprise. In some cases they have been considered legal documents and used as such in courts of law. There is an ethical expectation that scientists record their data honestly. Many times, working with my mentor, biologist Skip Snow in the Everglades National Park Python Project, I have seen Skip refer to previous entries when confronted with data that he thinks may provide a clue to a new line of investigation. Researchers don't leave home without notebooks.

WOrKING WITH ENGLISH LANGUAGE LEArNEr (ELL) POPULATIONS

Now, suppose that members of your class are from other cultures and have a limited knowledge of the English language. Of what use is inquiry science with such a population and how can you use the discipline to increase both their language learning and their science skills and knowledge?

First of all, let's take a look at the problems associated with learning with the handicap of limited language understanding. Lee (2005) in her summary of research on ELL students and science learning, points to the fact that students who are not from the dominant culture are not aware of the rules and norms of that culture. Some may come from cultures in which questioning (especially of elders) is not encouraged and where inquiry is not supported. Obviously, to help these children cross over from the culture of home to the culture of school, the rules and norms of the new culture must be explained carefully and visibly, and the students must be helped to take responsibility for their own learning. You can find specific help in a recent NSTA publication by Ann Fathman and David Crowther (2006) entitled *Science for English Language Learners: K–12 Classroom Strategies*. Also very helpful is another NSTA publication, *Linking Science and Literacy in the K–8 Classroom*, Chapter 12, "English Language Development and the Science-Literacy Connection"(Douglas and Worth 2006). Add to this array of written help two more books: *Teaching Science to English Language Learners: Building on Students' Strengths* (Rosebery and Warren 2008) and *Science for English Language Learners: K–12 Classroom Strategies* (Fathman and Crowther 2006). Finally, an article from *Science and Children* (Buck 2000) entitled "Teaching Science to English-as-Second Language Learners" has many useful suggestions for working with ELL students.

I can summarize as best as I can a few ideas and will also put them into the teacher background sections when appropriate.

Experts agree that vocabulary building is very important for ELL students. You can focus on helping these students identify objects they will be working with in their native language and in English. These words can be entered in science

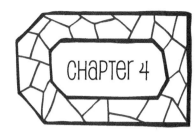
notebooks. Some teachers have been successful in using a teaching device called a "working word wall." This is an ongoing poster with graphics and words that are added to the poster as the unit progresses. When possible, real items or pictures are taped to the poster. This is visible for constant review and kept in a prominent location, since it is helpful for all students, not just the ELL students.

Many teachers suggest that the group work afforded by inquiry teaching helps ELL students understand the process and the content. Pairing ELL students with English speakers will facilitate learning since often students are more comfortable receiving help from peers than from the teacher. They are more likely to ask questions of peers as well. It is also likely that explanations from fellow students may be more helpful, since they'll probably explain things in language more suitable to those of their own age and development.

Use the chalkboard or whiteboard more often. Connect visuals with vocabulary words. Remember that science depends upon the language of discourse. You might also consider inviting parents into the classroom so that they can witness what you are doing to help their children to learn English and science. Spend more time focusing on the process of inquiry so that the ELL students will begin to understand how they can take control over their own learning and problem solving.

The SIOP model (Echevarria, Vogt, and Short 2000) has been earning popularity lately with teachers who are finding success in teaching science to ELL students. SIOP is an acronym for Sheltered Instruction Observation Protocol. It emphasizes hands-on/minds-on types of science activities that require ELL students to interact with their peers using academic English. You can reach the SIOP Institute website at *www.siopinstitute.net*. While it is difficult to summarize the model succinctly, the focus is on melding the use of academic language with inquiry-based instruction. Every opportunity to combine activity and inquiry should be taken and all of the many types of using language be stressed. This would include writing, speaking, listening, and reading. There is also a strong emphasis on ELL students being paired with competent English language speakers so that they can listen and practice using the vocabulary with those students who have a better command of the language.

In short, the difference between most other ESL programs and Sheltered Instruction is that in the latter, the emphasis is on connecting the content area learning and language learning in such a way that they enhance each other rather than focusing on either the content or the language learning as separate entities. In many programs it is assumed that ELL students cannot master the content of the various subjects because of their lack of language proficency. Sheltered Instruction assumes that given more opportunities to speak, write, read, talk and listen in the context of any subject's language base, ELL students can master the content as well as the academic language that goes with the content.

Teachers also need to be more linguistically present during classroom management tasks. They need to talk with students to make sure they are interpreting their inquiry tasks and learning how to explain their observations and conclusions in their new language. The teacher's role includes making sure students are focused by reminding them to write things down and to help them discuss their findings in English. As I said before, it is not only the ELL students who need to work on their academic language but all students who need to learn that science has a way of using language and syntax that is different than other disciplines. All students can benefit from being considered Science Language Learners.

And now, on to the stories which I hope will inspire your students to become active inquirers and enjoy science as an everyday activity in their lives.

references

Buck, G. A. 2000. Teaching science to English-as-second language learners. *Science and Children* 38 (3): 38–41.

Calkins, L. M. 2000. *The art of teching reading.* Boston: Allyn and Bacon.

Campbell, B., and L. Fulton. 2003. *Science notebooks: Writing about inquiry.* Portsmouth, NH: Heinemann.

Cervetti, G. N., P. D. Pearson, M. Bravo, and J. Barber. 2006. Reading and writing in the service of inquiry-based science. In *Linking science and literacy in the K–8 classroom,* ed. R. Douglas, and K. Worth, 221–244. Arlington, VA: NSTA Press.

Douglas, R., and K. Worth, eds. 2006. *Linking science and literacy in the K–8 classroom.* Arlington, VA: NSTA Press.

Echevarria, J., M. E. Vogt, and D. Short. 2000. *Making content comprehensible for English language learners: The SIOP model.* Needham Heights. MA: Allyn and Bacon.

Fathman, A., and D. Crowther. 2006. *Science for English language learners: K–12 classroom strategies.* Arlington, VA: NSTA Press.

Gee, J. P. 2004. Language in the science classroom: Academic social languages as the heart of school-based literacy. In *Crossing borders in literacy and science instruction: Perspectives on theory and practice,* ed. E. W. Saul, 13–32. Newark, International Reading Association.

Joyce, J. 1922. *Ulysses.* Repr., New York: Vintage, 1990. Page reference is to the 1990 edition.

Keeley, P., F. Eberle, and C. Dorsey. 2008. *Uncovering student ideas in science, volume 3: Another 25 formative assessment probes.* Arlington, VA: NSTA Press.

Keeley, P., F. Eberle, and L. Farrin. 2005. *Uncovering student ideas in science, volume 1: 25 formative assessment probes.* Arlington, VA: NSTA Press.

Keeley, P., F. Eberle, and J. Tugel. 2007. *Uncovering student ideas in science, volume 2: 25 more formative assessment probes.* Arlington, VA: NSTA Press.

Keeley, P., and J. Tugel. 2009. *Uncovering student ideas in science, volume 4: 25 new formative assessment probes.* Arlington, VA: NSTA Press.

Lee, O. 2005. Science education and student diversity: Summary of synthesis and research agenda. *Journal of Education for Students Placed At Risk* 10 (4): 431–440.

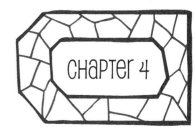

Padilla M. J., K. D. Muth, and R. K. Padilla. 1991. Science and reading: Many process skills in common? In *Science learning: Processes and applications*, ed. C. M. Santa and D. E. Alvermann, 14–19. Newark, DE: International Reading Association.

Palincsar, A. S., and S. J. Magnusson. 2001. The interplay of firsthand and text-based investigations to model and support the development of scientific knowledge and reasoning. In *Cognition and instruction: Twenty-five years of progress*, ed. S. Carver and D. Klahr, 151–194. Mahwah, NJ: Lawrence Erlbaum.

Postman, N. 1979. *Teaching as a conserving activity*. New York: Delacorte.

Reiss, M. J. 2002. Reforming school science education in the light of pupil views and the boundaries of science. *School Science Review* 84 (307).

Rosebery, A. S., and B. Warren, Eds. 2008. *Teaching science to English language learners: Building on students' strengths.* Arlington, VA: NSTA Press.

Winokur, J., and K. Worth. 2006. Talk in the science classroom: Looking at what students and teachers need to know and be able to do. In *Linking science and literacy in the K–8 classroom,* ed. R. Douglas and K. Worth, 43–58. Arlington, VA: NSTA Press.

THE STORIES AND BACKGROUND MATERIALS FOR TEACHERS

everyday earth and space science mysteries matrix

Stories

Basic Concepts	Moon Tricks	Where Are the Acorns?	Master Gardener	Frosty Morning	The Little Tent That Cried
States of Matter			X	X	X
Change of State			X	X	X
Physical Change			X	X	X
Melting			X	X	
Systems	X	X	X	X	X
Light	X	X			
Reflection	X	X		X	
Heat Energy			X	X	X
Temperature				X	X
Energy			X	X	X
Water Cycle				X	X
Rock Cycle			X		
Evaporation				X	X
Condensation				X	X
Weathering			X		
Erosion			X		
Deposition			X		
Rotation/Revolution	X	X			
Moon Phases	X				
Time	X	X			

Basic Concepts	What's Hiding in the Woodpile?	The New Greenhouse	Rotten Apples	Now Just Wait a Minute!	Cool It, Dude!
States of Matter					X
Phase Change					X
Heat Energy	X	X			X
Physical Change					X
Energy Spectrum		X			
Temperature		X			X
Conservation of Matter		X	X		X
Living Things			X		
Nature of Technology				X	
Design and Systems				X	
Gravity				X	
Time				X	
Technological Design				X	
Flow of Energy	X				X
Recycling Matter			X		
Weather and Climate		X			
Light		X			

continued

Basic Concepts	Where Did the Puddles Go?	What Are the Chances?	Here's the Crusher	Daylight Saving Time	A Day on Bare Mountain
States of Matter	X				X
Phase Change	X		X		X
Heat Energy	X				X
Physical Change			X	X	X
Energy Spectrum			X	X	
Temperature	X		X	X	X
Conservation of Matter	X	X	X	X	X
Nature of Technology				X	
Design and Systems		X	X	X	
Gravity		X			X
Time				X	
Technological Design				X	
Flow of Energy	X		X		X
Recycling Matter	X		X		
Weather and Climate	X	X			X
Light	X			X	

Basic Concepts	The Coldest Time	Is the Earth Getting Heavier?	What's the Moon Like Around the World?	Sunrise, Sunset
Solar Energy	X			
Temperature	X			
Heat	X			
Radiational Cooling	X			
Weather	X			
Climate	X			
Recycling of Matter		X		
Decay		X		
Decomposition		X		
Conservation of Matter		X		
Closed System	X	X		
Equinox				X
Solstice			X	X
Latitude			X	X
Earth's Tilt				X
Reflection			X	
Revolution			X	X
Moon Phases			X	
Earth-Moon-Sun System			X	X

CHAPTER 5
MOON TRICKS

Frankie turned eight on April 2nd. For a birthday present, he got a new bicycle—a 16-speed trail bike so that he could go out on the trails with his mom, dad, and older sisters, Karen and Martha. But his best present was a room of his own. His parents had just bought a new house in the same neighborhood and the new house had more rooms. His new room had windows looking out over the backyard. His mom told him they faced east.

The first day in the new house was fun. All of the furniture from the old house had to be put into new places and it was like putting together a puzzle. For their first meal in the new house, they sent out for pizza. Everybody was laughing and having a great time. But finally it was bedtime for Frankie. His mom and dad went up with him to his room to tuck him in. Frankie had placed his bed so that when he was lying in it, he could look across the room and out of some windows up near the ceiling. After he got into his pajamas, crawled into bed, and said goodnight to everyone, his parents turned out the light.

"Wow! Look at that!" said Frankie.

"Look at what?" mom asked as she stood in the dark room.

"The Moon! It's right there in the middle of my window! It's like a picture with a frame around it."

"So it is," said Dad. "Lucky you to have a room with the Moon looking in your window at bedtime each night."

The Moon was full—a big white circle, perfectly round and beaming light into Frankie's room. Frankie went to sleep easily. Even being in a strange room in a strange house seemed less scary since he had the Moon as his own night-light.

The next night, Frankie went up to bed and looked forward to seeing his new friend the Moon shining in his window. After the goodnights, the light was turned off and Frankie looked over at the window.

"Hey!" he shouted. "No Moon! Where is it?"

He bounded out of bed and looked up at the clear starry sky and saw no Moon at all. He felt cheated. The next night, no Moon again.

Two nights later, still no Moon at bedtime. What a disappointment! Bedtime wasn't as cool as it was on his first night.

A few days later, Frankie awoke in the middle of the night. A police car whizzed by the house, sirens screaming. Frankie sat up, frightened by the noise. He looked over and saw, to his amazement, his old friend the Moon, framed in the window. It was the Moon all right but it was not big and round. It looked like someone had cut off the right half of the circle.

Frankie was puzzled by this but too tired to think about it. It was three o'clock in the morning after all and he quickly dropped off to sleep again.

The next morning at breakfast he remembered what happened the night before and told everyone at the breakfast table what he had seen. Everyone seemed to have a different opinion.

Karen said there were clouds covering half of the Moon. She didn't know why it was outside his window at three o'clock. She also said Frankie might have been dreaming.

Martha thought that the Moon changed shapes through the night. She said it came up full and by the time it set, it was just a sliver of light.

Everyone laughed when Mom said she was pretty sure she sometimes saw the Moon in the daytime. But Dad agreed and said that maybe the Moon rose at different times each day. But that didn't explain the different shape.

Frankie was still left with a puzzle because nobody was sure of anything. Frankie wanted to be able to predict where the Moon would be at certain times and what shape it would have. How can you and Frankie find out why the Moon was showing up framed in his window at different times and in different shapes?

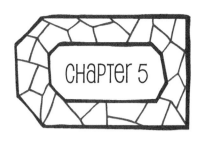

PURPOSE

The apparent daily motion of the Moon and other celestial objects through the sky is a major science concept. This story is designed to call attention to the changes in position and shape of the Moon over time. It is part of a larger conceptual scheme known as periodic motion. Everything is in motion, and finding patterns that repeat themselves is one of the hallmarks of science. Other examples of periodic motion are pendulums, seasons, and sound frequencies.

The Moon is such a common object in the sky that it seems a paradox that the majority of the population so poorly understands this familiar object. It is probably the fact that it is so familiar that it is taken for granted, like the number of steps you descend as you leave your home each day or the color of your best friend's eyes. Therefore, the purpose of the story is to motivate students to observe the Moon each day, record their observations and find the patterns in the Moon's movement and shape. For older students, the reasons for these patterns might also be the curricular goal.

RELATED CONCEPTS

- Periodic motion
- Time
- Reflection
- Revolution
- Astronomical motion
- Pattern seeking
- Light
- Rotation

DON'T BE SURPRISED

Your students may well echo the comments given by the story's characters. Since you will be recording these ideas, they will constitute questions for your evolving curriculum. The most common misconception regarding Moon phases is that the shadow of the Earth causes the phases. Some students believe that planets other than the Earth cast shadows on the Moon. Clouds or other planets are often blamed for blotting out parts of the Moon and some believe like Frankie's sister that the Moon undergoes a complete phase change in one day. Research shows that these ideas are common throughout the world and often persist into adulthood. Overcoming these ideas requires that students judge for themselves that the ideas listed above are not plausible, necessary, or not even observable. In order to combat the Earth shadow misconception, it is necessary for the students to experience a model that shows that the Earth does not enter into phase change and is responsible only for Moon eclipses. Most of these misconceptions are those voiced by some members of Frankie's family in the story. In this story Mom and Dad both made valid comments at breakfast.

CONTENT BACKGROUND

The first thing you will want to do is to keep a Moon journal yourself, as you might expect your students to do. You really should do this before you begin to teach this concept since it will prepare you for what your students will be bringing to class each day. (See the sections Using the Story With Grades K–4 and Using the Story With Grades 5–8 before you begin.) You will find that each day the Moon will appear to rise later than the day before. Since your students will be doing the same thing, you can expect the following data to be brought into your classroom each night the Moon can be seen, but of course you can expect that there will be gaps due to inclement weather.

If you began as suggested, on the night after the new Moon, you would find the Moon as a tiny, brilliantly lit crescent shape in the west, just before sunset. As the days pass, you would find that each day, the crescent would become more filled out until in about a week, it would be shaped like a half sphere with the curved side toward the setting Sun. In the week that follows the "first quarter Moon," the Moon's lighted portion would continue to grow so that about two weeks after the first crescent appeared, the Moon would appear to rise in the east as a fully lighted sphere, the full Moon. By this time, you would also notice that the full Moon rises at almost the same time as the Sun is setting. This observation would appear to suggest that the Moon and Sun would have to be on opposite sides of the Earth at this juncture. If the Earth's rotation causes the apparent rising and setting of the Sun and Moon, then the Sun's setting coinciding with the Moon's rising would best be explained by the Sun and Moon being at opposition with the Earth in the middle. Since the Moon's orbit is tilted slightly from the plane of the Earth's rotation, barring eclipses, the Sun would be shining directly on the Moon's surface showing all of it to observers on the Earth. Thus, a full Moon! This will appear as an important pattern since the Moon will have risen later each day. The major finding would be that you would have seen the Moon setting in the west just after sunset on day 1 of your observation and rising at sunset on approximately the 14th day of your observation.

By the night after the full Moon, the moonrise might be past your bedtime and that of your students. No Moon? The next day you see the Moon up there almost full in the daytime! This may come as a surprise to many that the Moon can be seen in the daytime. As the time progresses to week 3, the Moon is still visible in the daytime and very early in the morning and the curve of the lighted part of the Moon points in the direction of the sunrise! This is the Moon that Frankie saw at 3:00 a.m. when awakened by the noises outside his window. By the end of week four, the Moon when visible is a crescent again but this time the curve is the opposite of what you saw on the first night. The next night you see no Moon at all, day or night and this is the "new Moon." Then the cycle starts all over again. It is important here to help the students realize that the light that falls on the Moon is reflected light from the Sun and that the Moon does not have a light source of its own.

You and your students would also note that if you observed the Moon at a specific time each night, the Moon would appear a bit more toward the east in the

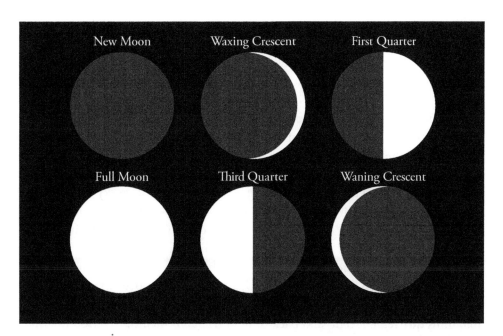

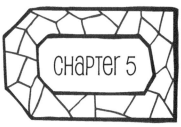

sky as it is observed from your vantage point. It was observed farther and farther east until it finally rises in its "full" phase.

Your young students might well make a picture graph of their drawings and the shapes of the Moon over the period of time they observed. They can see the pattern of crescent to first quarter to full to third quarter to new to crescent again and so on. This satisfies the expectation of the early grade standards.

If, as suggested in the "how to use the story" sections, you gave your students a probe or had them list the "things we know—our best thinking so far," on a large sheet of paper, the ideas that no longer hold water can be eliminated one by one as the evidence shows to the contrary.

The Moon revolves around the Earth approximately once every 28 days and appears to rise above the eastern horizon each day, moving across the sky to its setting point on the western horizon. Observers on Earth see this because the Earth rotates from west to east once each 24 hours and as the Earth reaches the position where the Moon is visible in the east, the Moon is said to "rise." As the Earth continues to rotate, the Moon appears to move across the sky and finally to "set" or disappear below the horizon. In the meantime, the Moon also revolves around the Earth, in the same direction as the Earth rotates on its axis, thus moving ahead, relatively speaking, of any 24-hour observation point on the rotating Earth. In other words, the observer sees the moonrise on day 1, at a given time. At the next rising of the Moon, (day 2) the Moon will have moved 1/28th of its revolution around the Earth during the elapsed 24 hours. Thus, the observer will have to wait until the Earth has moved a bit more than one revolution (24 hours) in order to catch up to the moonrise. This will result in a moonrise that will be a bit later each day. The average yearly difference is about 50 minutes per day but this varies from day to day and season to season.

Frankie was able to see the full Moon in his window by pure luck on his moving day. He expected to see it there night after night but this was not to be. First, the Moon rose later on the next night and on subsequent nights so that it was not visible at his bedtime. Second, due to the position of the Moon in relation to the Sun and the Earth, the Moon's lighted surface was visible from a different spatial vantage point from Earth and appeared to change shape from full to partial circle with its right half in darkness. This is referred to as the third quarter. Contrary to many beliefs, the Moon at this point is visible mostly during the a.m. from early predawn hours and the normal daylight hours to its setting before sunset. Then as the cycle progresses, a tiny crescent will be noticed in the western sky shortly before sunset and the cycle will repeat. Each day the Moon will rise later and more of the lighted surface visible from the Earth will become visible until the full Moon, which Frankie saw in his window. Then the lighted surface will begin to appear smaller until the new Moon, which is not visible since the lighted surface of the Moon is hidden from viewers on the Earth. So in summary, over an approximately 28-day period, the Moon's phases go from a new Moon, a crescent, a first quarter, full, third quarter, crescent, and back to new. (Actually, half of the Moon's surface is always illuminated by sunlight but due to our relative position to the Moon we may see only part of that illuminated surface.) Go to the following website for a demonstration: *www.noao.edu/education/phases/phases_demo.html.*

If you adhere to the National Standards and the Benchmarks in your curriculum development, you will want to consider the following:

- (1) Grades K–4 children will want to focus on the **patterns** in the shape of the Moon changes from day to day. They can also look at the rising and setting times of the Moon for added information to solve the mystery.

- (2) Grades 6–8 children will be guided into looking into the pattern and also into the **reasons** for the apparent changes in the shape and timing of the Moon cycle, thus gathering evidence for solving the mystery.

related ideas from the national science education standards (NrC 1996)

K–4: Changes in Earth and Sky
- Objects in the sky have patterns of movement. The observable shape of the Moon changes from day to day in a cycle that lasts about a month.

5–8: Earth in the Solar System
- Most objects in the Solar System are in regular and predictable motion. Those motions explain such phenomena as the day, the year, phases of the Moon, and eclipses.

related ideas from Benchmarks for Science Literacy (aaas 1993)

K–2: The Universe
- The Moon looks a little different every day, but looks the same again about every four weeks.

6–8: The Earth
- The Moon's orbit around the earth once in about 28 days changes what part of the Moon is lighted by the Sun and how much of that part can be seen from the Earth—the phases of the Moon.

USING THE STORY WITH GRADES K–4

Reading the story to the children, or having the children read the story themselves, is a matter of choice. A child's listening vocabulary is often greater than her reading vocabulary. Children enjoy being read a story and the reader can dramatize or emphasize certain parts of the story for effect. Since the story has no real ending, the children are asked to discuss the situation posed by the mystery. "What do we need to know in order to write an ending to this story?" There will be many ideas suggested, and we have found that the best way to use them is to record them on a large piece of paper, labeled, "Our Best Thinking Until Now." All ideas are accepted. Some may agree or disagree with various members of the family in the story, while others may suggest new ideas. When these ideas are posted, the reader, referring to the title of the page, suggests that each of these ideas or statements can be turned into a question. For example, a posted idea may be "Clouds cover the Moon and cause it to change shape." This can be changed to "Do we have any evidence that clouds cover the Moon and cause it to change shape?"

After this has been accomplished, the teacher asks the children how we can gather evidence to support or disprove the questions. The children usually respond by saying that they can "look it up" or with a little guidance, they can keep a Moon record for a month to see what they can find out. This leads to keeping a Moon journal in which the students identify a spot where they can observe the sky. The next question becomes, "What shall we record?" Here the leader can go back through the story and point out what kinds of things they have to observe in order to find the answer to the mystery. With guidance, the children will decide that the journal should include the time of the observation, the position of the Moon in the sky, how high it is in the sky and its shape. If the class can see the importance of keeping similar records, they can agree on a time (appropriate for their bedtime) and a method of recording so comparisons can be made when they share their journals.

Sending a letter home to parents explaining the project and asking them to help their child in remembering the time for observation and helping them to record all of the required data will give you some help. Students and parents will have to be made aware of the term *horizon* and its location even if mountains are present. It should be a point directly ahead of them where the sky meets the land. In mountainous regions, this would be at the end of the hand held out parallel to flat ground. The height of the Moon can be measured simply by using the "fist" method. Hold the fist of one hand, arm held parallel to the ground, at the horizon and then place the fist of the other hand directly upon the first fist. Holding that fist steady, move the other fist on that one and so on until the last fist covers the Moon. Count the number of fists and record this as data. Children of any one age group will have fists so close in size that little difference will be found among children to confuse the data.

The position of the Moon, as observed at the chosen time each day, will be important too, as it will move from west to east. We suggest that the students draw a landmark on their data sheet (a natural landmark or one placed on the horizon). Then students can mark the position of the Moon in relation to this landmark. It could be a tree, a house, a mound of earth, a bush, or a barrel. It is a point of reference since it never moves. Ask the children if it is important for them to observe from the same place each time. Discuss this with them.

Finally, the timing of the beginning of the data collection is of utmost importance.

It should be started the night or possibly on two nights following the *new* Moon. The reason for this is that the Moon will be visible before they go to bed. We suggest a time just after sunset in your location. During the autumn this should be approximately 7:00 p.m. Be sure to be aware of time changes due to Daylight Saving Time if this is prolonged to a later date. In order to know when sunsets and Moon settings and risings occur, you will need a local almanac. *The Old Farmer's Almanac* is usually available at supermarkets or bookstores. There is usually one published for all areas of the country. Otherwise, many newspapers carry an almanac as well. You can also look on the internet for these times at *www.almanac.com*. This is extremely important because if you were to start this journal and data collection at full Moon time, the next risings of the Moon would become later and later and well past your students' bedtimes. With data collection beginning at first crescent, there are ample opportunities for viewing the Moon until its full phase. After that, until the new Moon, you can most often find the Moon during the daytime hours. Your almanac will help you with this.

Each day you can record the findings on a large chart so that the entire cycle can be seen. Weather may well interfere, with cloudy nights. If you can spare two months for observation this will usually provide enough data to complete a cycle. Obviously, you can carry out the Moon journal recording while engaged in another unit of study since it takes only a small part of the school day.

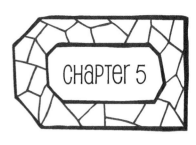

Finishing the story becomes apparent when enough data are collected that the children realize that the Moon rising later each night kept Frankie from having his "nightlight" each night. Had he gone to bed later each night, he might have had the pleasure of some moonlight but that is usually not the case. Since Frankie saw the third quarter Moon a week later in the middle of the night, the later rising of the Moon and the phase change explains his observation. You can encourage the children to write the ending to the story as Frankie explains to his family what he has found that convinced him that the mystery is solved.

USING THE STORY WITH GRADES 5–8

Most of the same techniques suggested for the K–4 classes are valid for grades 5–8 as well. However, the emphasis may shift to the reasons for the phases than on the presence of the patterns. Remember that many students, including college postgraduates, may be ignorant of the patterns and their ideas must at least be reviewed, preferably through a Moon journal. In the case of older students, they are usually developmentally ready to pursue the mechanics behind the reasons for the phases. It is important to go through the steps of discussing the Moon journal data so that the students have the opportunity to voice their preconceptions and test them against reality. Then the demonstration of Moon phases will have much more meaning to them. Remember Duckworth's admonition that the student must have "done a major part of the work already, one has to have developed a network of ideas in which to imbed the [experiences]" (1986). Having discussed their findings and argued these data, the students will be more receptive to any demonstration. They may even have models of their own to try. This of course would be the best scenario but one must be prepared to provide a model for them to ponder if their models are not forthcoming or do not work.

Your older students are expected to explain this pattern and the obvious changes in the lighted parts of the Moon. As your students discuss their findings each day, they should be invited to manipulate three balls representing the Sun, Moon, and Earth in such a way that the observed Moon pattern (the reflected light patterns) could be reproduced. Having a floor lamp as a light source (Sun) might help or a bright light in the center of a darkened room would allow them to try different configurations of Sun, Moon, and Earth relationships. It will probably be noticed that in none of the configurations does the Earth's shadow need to fall on the Moon in order for them to witness a phase change. They might also notice that at the first quarter, there is no curve to the lighted portion, eliminating the misconception that it is the Earth's shadow that causes the phases. They will also see that it is impossible to create a gibbous Moon. They should have noticed that the clouds had nothing to do with the Moon's pattern from night to night and that the Moon does not go through the whole phase pattern, eliminating other common conceptions.

MODELS FOR EXPLAINING THE PHASES OF THE MOON AS SEEN FROM EARTH

There are many articles devoted to this demonstration of the Earth-Sun-Moon relationship. It is important when using a physical model utilizing the children directly to try to be as true to the actual astronomical situation as possible. My favorite is one where students work in pairs, one being the Earth and the other moving the "Moon" which is a Styrofoam ball impaled on a dowel. A light with a bare bulb is placed in the center of a darkened room at about two meters above the floor with the student pairs arranged around it. The student in each pair who is Earth views the Moon from the peak of "mount nose," and spins slowly one complete turn, while the other student moves the Moon around the Earth each time the Earth rotates, in the same direction that Earth rotates. The Earth observer will notice that she has to turn a bit more than 180° in order to see the Moon because it has progressed in its revolution around the Earth. This will explain the reason for the later rise of the Moon each night, and the phase changes will also be obvious as the Earth continues to rotate and the Moon continues to revolve. Partners then switch and repeat the exercise.

It is very important that the students have the opportunity to struggle with their own models and observe the Moon firsthand before being introduced to the model described above. They should be able to see that this model allows them to verify the things they have observed and resolves the puzzles remaining from their prior attempts to create a model.

In addition you may read "The Sky's the Limit" from *Science and Children*, September 1999; "Look to the Moon," from *Science and Children,* November/December 1996; or "The Moon Project" from *Science and Children,* March 2006. NASA has a website that describes a Moon phase demonstration at *http://education. jpl.nasa.gov/educators/moonphase.html*. Additional resources on the Earth and Moon are available at *www.fourmilab.ch/earthview/vplanet.html*.

references

Driver, R., A. Squires, P. Rushworth, and V. Wood-Robinson. 1994. *Making sense of secondary science: Research into children's ideas.* New York: Routledge Falmer.

Duckworth, E. 1986. *Inventing density.* Grand Forks, ND: Center for Teaching and Learning, University of North Dakota.

Foster, G. 1996. Look to the moon. *Science and Children* 34 (3): 30–33.

Roberts, D. 1999. The sky's the limit. *Science and Children* 37 (1): 33–37.

Trundle, K. C., S. Willmore, and W. S. Smith. 2006. The moon project. *Science and Children* 43 (7): 52–55.

Yankee Publishing. *The old farmer's almanac,* published yearly since 1792. Dublin, NH: Yankee Publishing.

CHAPTER 6

WHAT'S THE MOON LIKE AROUND THE WORLD?

aqil and Durrah had recently arrived in the United States from the Middle East and were enjoying the night sky in their new home in the "Middle West." The light from the Sun had disappeared, and the crescent-shaped Moon had just come into view low in the western sky. Aaqil began to wonder out loud.

"I think that the Moon is very beautiful, Durrah. I love this shape that looks like a curved circle. I wonder what the Moon looks like tonight in our home country?"

"Yes, it looks very beautiful, and I wonder, too, what it looks like way down south in Australia. They say everything is backward and upside down in the country they call 'Down Under,'" she replied.

"I think," said Aaqil, "They call it 'Down Under' because it is way south of the equator. Our teacher says that compared to us, people down there are standing upside down. I suppose they see everything like they were standing on their head!"

"So, if we have a full Moon here, they would have a new Moon and see nothing in the sky, right?"

"I'm not so sure about that, Durrah," said her brother. "I think maybe the moon is the same everywhere on Earth, but I don't really know."

"Well, our country across the ocean is a long way from here, even though we are both in the northern part of the world. So how can the Moon be the same shape?"

"It is the same Moon, isn't it?"

"Yes, but it is so far away from what we are seeing that it can't be the same shape!"

"There must be a way to find out."

"I know! We can email our friends we left behind. They would answer our question about what the Moon looks like from east to west," said Durrah. "But what about north to south?"

"Let's try to find somebody who lives way down south."

NATIONAL SCIENCE TEACHERS ASSOCIATION

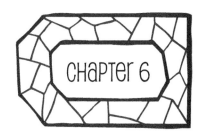
PURPOSE

A great deal of confusion arises from the lack of understanding about the Moon's journey around the Earth and its apparent shape changes. This investigation is aimed at confronting this confusion by looking at the Earth-Moon system and how it appears from various points on the Earth. Since the discussion includes the two hemispheres, children will be able to see that all of the phases of the Moon are opposites in every way.

RELATED CONCEPTS

- Reflection
- Earth-Moon-Sun relationship
- Moon phases
- Orbits
- Systems

DON'T BE SURPRISED

The causes of Moon phases are an enigma to many children, as well as adults. Witness the responses given by the graduates of Harvard University and students in a nearby high school in the film *A Private Universe* (Schneps 1987). Things haven't changed a lot, from my experience, in the last 24 years, and it is likely that your students will have an interesting time discussing and arguing the points made in the story. Many students and adults believe that the shape of the moon as it changes phases is due to the shadow of the Earth on the Moon. It is indeed one of the strongest astronomical misconceptions.

CONTENT BACKGROUND

Imagine a situation where the Moon's phases are different all across the world. Now imagine you have to prepare a calendar for sale around the world with these Moon phases pictured each month. What a mess! It's a good thing this is not the case. The Moon looks pretty much the same no matter where you live *unless* you live in the southern hemisphere. If you live far enough "down under" you are literally standing on your head compared to those who live in the northern hemisphere, so you must see the shape of the Moon differently. However, anywhere on the Earth we see the sunlit portion of the Moon, and depending upon our position in regard to the Earth and Moon, that sunlit portion of the Moon will appear different each day. This means that it is impossible for some of us to see a full Moon and others to see a new Moon.

If you live in the United States and see a full Moon, what do you expect a friend living in Chile, South America to see? In volume 1 of Page Keeley's *Uncovering Student Ideas in Science* (2005) there is a probe called "Gazing at the Moon"

that asks this very question. Many people believe that because of the hemispheric difference, the difference in Moon phases is opposite. This is true, but it does not mean that a full Moon in the northern hemisphere becomes a new Moon in the southern hemisphere. Rather, it means that the crescent shape is pointing in the opposite direction.

This difference will challenge your spatial relationships to their limits. Imagine that you are looking at the Moon in its crescent phase. Now, imagine that you are looking at the same image while standing on your head. The horns of the crescent would be pointing in the opposite direction, right? Try it and see. If you can't stand on your head just bend over until you are looking upside down.

The Earth rotates in a west-to-east direction so the Moon will rise in the East and set in the West regardless of where you live. However, if you live in the southern hemisphere (SH) the Moon will appear to travel across the northern sky rather than in the southern sky as it does in the northern hemisphere (NH). In the NH, the lighted portion of the Moon grows from right to left over the 29.5 days of its full cycle. In the SH the lighted portion of the Moon grows from left to right.

Because of this, the rule we have about the waxing and waning Moon is opposite in the NH and SH. In the NH we say that the Moon is waxing (moving toward full phase) if the shape of the first quarter looks like a growing C and the last quarter looks like a D. The opposite is true in the SH. I found that I needed some help on this so I used a website to visualize the differences. One good one is *www.woodlands-junior.kent.sch.uk/time/moon/phases.html* (Barrow 2008). Scroll down to the bottom and click on "Moon Around the World." It also mentions that near the equator, the crescent moon looks either like a U or a smile when it sets. This is because the Moon's orbit is very close to that of Earth's orbit and so is at less of an angle than it is in either the far NH or SH.

For those of you who are beginning to zone out and say this is much too difficult for your students, I recommend you read "Meeting the Moon From a Global Perspective" (Smith 2003). In this article, the author chronicles a project that connected students from all parts of the globe exchanging their views of the moon and how it related to their global position and their cultures. I wish I could have been part of that project and posit that, with the suggestions and caveats listed in the article, it is possible to re-create much of the same excitement and learning on your own. There are privacy concerns involved with children corresponding with each other, but these problems are addressed in the article. The gains in cultural understanding and science knowledge far overshadowed the difficulties in setting this project up. In fact, if you use your favorite search engine and type in "current global school science connections," you may be able to find some ongoing projects that may interest you and your school system.

NATIONAL SCIENCE TEACHERS ASSOCIATION

related IDeas FROM THe national science education standards (NRC 1996)

K–4 Changes in Earth and Sky
- Objects in the sky have patterns of movement. The observable shape of the Moon changes from day to day in a cycle that lasts about a month.

5–8 Earth in the Solar System
- Most objects in the solar system are in regular and predictable motion. Those motions explain such phenomena as the day, the year, phases of the Moon, and eclipses.

related IDeas FROM BeNCHMaRKS FOR science Literacy (aaas 1993)

K–2: The Universe
- The Moon looks a little different every day, but looks the same again about every four weeks.

3–5: The Universe
- The Earth is one of several planets that orbit the Sun, and the Moon orbits around the Earth.

6–8: The Earth
- The Moon's orbit around the Earth once in about 28 days changes what part of the Moon is lighted by the Sun and how much of that part can be seen from the Earth—the phases of the Moon.

USING THE STORY WITH GraDes K–4

Using this story with younger children will raise more questions about the phase changes in the Moon than anything else. Children with developing spatial skills will have a difficult time imagining people "standing upside down," in the southern hemisphere. The story, "Moon Tricks," in *Everyday Science Mysteries* (Konicek-Moran 2008) tells the story of a boy who moves into a new house and sees the full Moon in his bedroom window on his first night. He is disappointed on the second night to find the full Moon missing and awakes in the middle of the night a week later to see the Moon outlined in his window but in a different shape. The story

goes on to lead the students to observe the Moon over a period of time to see for themselves the pattern of phases that occur in the sky. Materials in *Everyday Science Mysteries* give elaborate directions for helping students keep Moon journals and use them to build models of the Earth-Moon system.

Even if you do not have access to this book, you can make arrangements with parents to help your students keep track of the Moon's direction, shape, and height above the horizon at a given time each day. This will result in a journal that can be discussed in class and give your students some firsthand data to analyze. Our experience is that children are not aware of these changes and will benefit greatly by observing them on a regular basis and collecting data about what they see.

USING THE STORY WITH GRADES 5–8

If your students in the middle grades are not aware of the reasons for the Moon's apparent changes in shape over the monthly period, I suggest the same story and background material offered to grades K–4 teachers. It is important that your students understand the celestial mechanisms that cause the pattern of the phases of the Moon before they can visualize the way others in the world see the Moon.

Your students should work in pairs in a darkened room with a single lamp about two meters above the floor acting as the Sun. One student will act as the Moon using a Styrofoam ball on a dowel. The other student is the Earth and observes the Moon as it moves around this student slowly. The Earth rotates once (the child spins once, back to the original place) and the Moon moves in the same direction so that the phases are seen as they go through the month of 29 days. Partners switch and the exercise is repeated.

Once the students have seen how this works, they should be able to repeat the activity and try to visualize the phase changes by bending over, assuming a modified upside-down position and seeing that the pattern is reversed. It may take several tries before the reverse pattern is seen by all of the students. Have the students acting as the Earth vary their positions so that they can see that they must be very far down in the SH in order to see the reverse pattern. At this point, they should also be asked if they think the Moon would look any different to an observer in the same hemisphere.

They should notice that the lighted area of the Moon is always the same regardless of the position of the observer, but that the observer's view is dependent upon the rotation of the Earth. Whether in France, Asia, or anywhere else on the globe, the Earth has to turn sufficiently so that Moon is visible, and of course since the Earth rotates from west to east, the farther west one lives, the later the Moon becomes visible. However, the lighted surface of the Moon is always the same shape regardless of hemisphere (i.e., crescent, quarter moon, full, and new). So Aaqil and Durrah in the story can be certain that their friends back home will be seeing the same moon shape that they do but their friends will have seen it earlier. In fact if they could call their friends, they would find that the Moon had come into view about seven hours earlier over there.

58

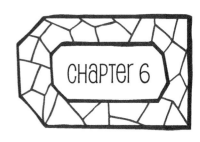

So, in essence, the Moon rises later each day but looks basically the same if all viewers are in the same hemisphere. The major change in views depends on being in different hemispheres. The SH and NH view the moon with a difference of 180°.

related books and NSTa Journal articles

Gilbert, S. W., and S. W. Ireton. 2003. *Understanding models in earth and space science*. Arlington, VA: NSTA Press.

Keeley, P., and J. Tugel. 2009. *Uncovering student ideas in science, volume 4: 25 new formative assessment probes*. Arlington, VA: NSTA Press.

Konicek-Moran, R. 2009. *More everyday science mysteries: Stories for inquiry-based science teaching*. Arlington, VA: NSTA Press.

Oates-Brockenstedt, C., and M. Oates. 2008. *Earth science success: 50 lesson plans for grades 6–9*. Arlington, VA: NSTA Press.

references

American Association for the Advancement of Science (AAAS).1993. *Benchmarks for science literacy*. New York: Oxford University Press.

Barrow, M. 2008. The phases of the moon. Woodlands Junior School, Kent. UK. *www.woodlands-junior.kent.sch.uk/time/moon/phases.html*

Keeley, P., F. Eberle, and L. Farrin. 2005. Gazing at the moon. In *Uncovering student ideas in science, volume 1: 25 formative assessment probes*, 177–181. Arlington, VA: NSTA Press.

Konicek-Moran, R. 2008. *Everyday science mysteries: Stories for inquiry-based science teaching*. Arlington, VA: NSTA Press.

National Research Council (NRC). 1996. *National science education standards*. Washington, DC: National Academies Press.

Schneps, M. 1986. *A private universe project*. Harvard Smithsonian Center for Astrophysics.

Smith, W. 2003. Meeting the moon from a global perspective. *Science Scope* 36 (8): 24–28.

CHAPTER 7
DAYLIGHT SAVING TIME

Tonight we set the clocks ahead one hour! It's the first Sunday in March tomorrow," Jackie said excitedly.

"What's the big deal about that?" said Denzel, her cousin who was visiting from the city. "We lose an hour of sleep 'cause we set the clocks ahead tonight and wake up tired."

"Well, we get an extra hour of daylight and besides that we save energy," said Jackie.

"What are you talking about? Messing with the clocks doesn't give us more daylight! And how do we save energy? Explain that to me," Denzel said, impatiently. Denzel could get irritated with his younger cousin but mostly he got along with her.

"Well," Jackie responded without a pause, "it's much brighter at supper time and the Sun sets later so we must be getting extra sunlight somehow. And that means we don't have to turn on our house lights so early and that saves energy!"

"Look," Denzel said, "you just messed with the clock, not the Sun. You ought to check and see if you get more Sun. And as far as the lights are concerned, we have to turn them on earlier in the morning so where's the savings on energy? We're just trading electricity in the morning for electricity in the evening."

"Anyway, I heard Grandpa say that some of his friends were afraid that the extra hours of sunlight would fry the plants, so there!"

"He was joking," laughed Denzel.

"No he wasn't," retorted Jackie. "And anyway how do you explain that everybody says we get our hour back when daylight saving time ends in November? If they didn't take the hour in the first place, how come we get it back?"

Denzel merely rolled his eyes.

"I think you'd better ask someone at your school about this, Jackie. Extra hours of daylight and saving energy seems pretty silly to me. But you take your ideas to school and talk it over. Let me know what happens."

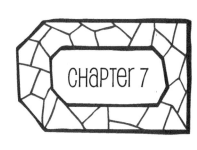

PURPOSE

Time is a difficult concept for many people to understand. We are dependent on our clocks, watches, and calendars. But before in our history, folks could pretty much tell the time just by looking at shadows or the position of the Sun in the sky. Even many adults do not realize that time is a construction made by civilizations to break up their days, months, and years for agricultural reasons. This story deals with the misconception that Jackie has about clocks setting time, while Denzel tries to make her understand that it is the relationship in the relative positions of the Earth and Sun that determines seasons, daylight and night, and other natural events important to humans—not clocks and calendars. Time and calendars were created by people after all because of their observations of the celestial bodies.

This story also provides an opportunity for children to debate the various questions surrounding the changing of clock time to save energy, help farmers, prevent traffic accidents, and prevent crime. All of the above have been mentioned as reasons for daylight saving time, yet there is still a great deal of controversy about the practice. It is a great opportunity for children to gather and evaluate data.

RELATED CONCEPTS

- Earth-Sun relationships
- Earth's motion
- Time
- Sun's apparent motion

DON'T BE SURPRISED

I have actually talked with adults who believe that the clocks we use are directly related to Sun time and that changing clocks will alter the Sun-Earth time relationship. It is not difficult to understand then that your students may have the same misconception. Little children usually begin thinking about time as connected to an event. Is it time to get up, go to bed, take a nap, or eat lunch? They are often upset when they want to watch their favorite TV show and are told that it doesn't come on until tomorrow. It should come on when they want it to! With this kind of start, it's no surprise when children do not realize that the origin of time was based on celestial events but created by humans, and that the various cultures have manipulated clocks, calendars, and holidays to fit their cultural needs and beliefs. The abstraction is tremendous; and the leap in thinking from time as an interval between holidays or meals to a celestial concept of time is quite a challenge. Perhaps this is why even adults are prone to misunderstanding such things as "spring forward and fall backward," not realizing that it is simply a mnemonic for what to do with clocks each spring and fall, not an actual reflection about what really happens with Sun time. The whole idea of daylight saving time is a mystery to many people.

CONTENT BACKGROUND

Time has probably always been tied to the celestial movements of the Sun, Moon, and several other planets. The ancient Greeks and Romans had Sun clocks and somehow decided to divide the day into 24 segments called hours. But due to the irregularities in the length of the day at various times of the year, some hours were longer than others. Civilizations that developed in the tropics had little seasonal differences, so their hours were quite regular. The farther north or south of the equator the civilization developed the more irregular their time systems. But since there was little need for very accurate time, they lived with these irregularities.

Archaeologists and historians are still trying to decipher the real reasons for places like Stonehenge in England and Chaco Canyon in New Mexico, and to understand the diverse calendars in Polynesia, Asia, and South and Central America. It is clear that the marking of time has always been of great interest to societies. Agriculture, for example, was a major concern for early civilizations, so indicating the times for planting and harvesting became paramount. Some civilizations, such as the Egyptians and Greeks, tried to find ways to measure more subtle time intervals as early as 1500 BC. These were fraught with problems.

It was not until ocean travelers were trying to measure their exact locations on the globe, especially longitude, that the need for very accurate timepieces became important. Then, of course, when commerce and travel had to rely on common forms of transportation and the economy became global, the world had to come together to agree upon a system of timekeeping that would make all sorts of industries and modes of transportation equivalent. Imagine how confusing it must have been when each city, state, and country was following its own time standards!

It was 1675 when most of the nations of the world accepted that time zones needed a starting place and agreed on Greenwich Mean Time. The Greenwich Observatory in England was set as longitude 0 degrees (despite French objections). Britain, in 1865, delineated time zones so trains could schedule arrival and departure times. Since local solar times are based on the Sun's position in the sky, the time is different depending upon where you happen to be. When transportation became faster and trains could cross entire countries in a short time, it became necessary to note that 12:00 (on a Sun clock) on the western edge of the country was a different time from 12:00 on the eastern edge. But it was not until 1929 that all nations agreed to the Coordinated Universal Time (UTC) that set up time zones for every 15 degrees of longitude around the globe, finally eliminating the confusion of every nation in the world adhering to its own time standards. Since the day is 24 hours long and a full circle contains 360 degrees, each time zone was limited to 15 degrees, an hour's difference. Within these parameters, some countries and provinces set up 30-minute differences for their internal time zones (e.g., Newfoundland, India, and Afghanistan) and some even use 15-minute differences. Other countries, such as China, use only one time zone even though their vast east and west boundaries far exceed 15 degrees. There is a great world

time zone map at *www.worldtimezone.com* if you would like to see how the world's time is divvied up.

Even though the idea of daylight saving time has been discussed since the 1700s, it was not accepted worldwide until 1918, and even then it was not acknowledged by farmers, especially in Indiana. There were those who still believed that it was an abomination because it provided extra sunlight. But it was promoted by those who stood to gain monetarily from its enactment, like sporting goods manufacturers, since it gave leisure time during the daylight hours to people who ended their working day at 5:00 p.m. Issues of traffic and pedestrian safety, energy savings, and crime prevention have often been given as reasons to promote the idea. Hawaii and Arizona are two states that still do not agree to use daylight saving time, along with over 40 countries in the world. Many have tried it for a few years and then decided not to use it. The main reason for rejection is the desire to stay on the same time schedule to trade with countries that do not use it.

Since the sunrise is one of the main indicators of the beginning of each work-day, we can see that as the Earth turns from west to east, it takes longer for the sunlight to reach the western edges of any time zone. If you look at an almanac with sunrise tables, you will notice that sunrise will be later for each distance far-ther to the west.

Let us say that you are traveling from Boston to Seattle on a nonstop flight that takes five hours and leaves at 8:00 a.m.. When you reach Seattle, you will find that only two hours have passed on the clock because even though you have been traveling for five hours, you have passed over three time zones to the west, subtracting three hours from the clock time. It is only 10:00 a.m. in Seattle. When you travel back to Boston, the same thing happens in reverse: five hours in the air, yet eight hours later when you arrive. You have passed over three time zones—this time to the east, which *adds* three hours on the clock time. Your body time will be on Seattle time, but your Boston clock will be later. This is the so-called jet lag—the difference experienced in your body between your body time and the time zone in which you arrive. This lag is more pronounced when you travel lon-ger distances. Our bodies have an "internal body clock," which develops over time when we live in one time zone. We go to sleep and wake on this internal clock. When we experience a different country's time, we must adjust to this, which may take several days. A good example of animal internal clocks can be seen when, as daylight saving time changes, dogs and cats beg for their meals at the usual time, not at the clock time. Your own stomach may begin to growl, even though the clock now says that you have an hour to wait.

Time is an interesting, yet difficult, abstraction for adults and children to comprehend. All of the experiences you can provide your students will help them understand what a fascinating topic it is. I love movies and books that deal with time travel, like the classic H.G. Wells book *The Time Machine*. Backward or for-ward, it makes no difference. It makes me wonder what it would be like to travel to another age.

related ideas from the National Science education Standards (NRC 1996)

K–4: *Objects in the Sky*

- The Sun, Moon, Stars, clouds, birds, and airplanes all have properties, locations, and movements that can be observed.

K–4: *Changes in Earth and Sky*

- Objects in the sky have patterns of movement. The Sun for example appears to move across the sky in the same way every day. But its path changes slowly over the seasons.

5–8: *Earth in the Solar System*

- Most objects in the solar system are in regular and predictable motion. These motions explain such phenomena as the day, the year, phases of the Moon, and eclipses.

related ideas from Benchmarks for Science Literacy (AAAS 1993)

K–2: *The Universe*

- The Sun can be seen only in the daytime but the Moon can be seen sometimes at night and sometimes during the day. The Sun, Moon, and stars all appear to move slowly across the sky.

K–2: *The Earth*

- Like all planets and stars, the Earth is approximately spherical in shape. The rotation of the Earth on its axis every 24 hours produces the day/night cycle. To people on Earth, this turning of the planet makes it seem as though the Sun, Moon, Planets, and stars are orbiting the Earth once a day.

6–8: *The Earth*

- Because the Earth turns daily on an axis that is tilted relative to the plane of the Earth's yearly orbit around the Sun, sunlight falls more intensely on different parts of Earth during the year.

USING THE STORY WITH GRADES K–4 AND GRADES 5–8

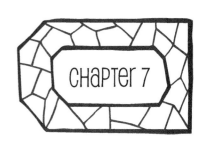

I am combining the grade-level ideas in this chapter because the activities even in a modified form may be appropriate for more than one level. If you are inclined to use this story to explore the nature of the motion of the Sun and Moon, please revisit the introduction to this book for case studies about how two teachers used a similar story at different grade levels. You will be able to see the yearlong process as these two teachers used the story "Where are the Acorns?" and find areas of inquiry in common with this story. I might also recommend that you look at a copy of *Everyday Science Mysteries* (Konicek-Moran 2008) pages 45–50, to see what other suggestions of an astronomical nature you might use. See "Moon Tricks" (Chapter 5 in this book) as well.

If you do not need to venture into the area of astronomy, I still suggest that you find out what your students know about the apparent movement of the Sun during the year. Without some knowledge of the importance of observations of the Sun during the day and over the year, the true meaning of time will be lost. You may want to set up a sundial in the correct direction of north and have the children read it and keep records of the Sun's motion each day. They will notice that the shadow that marks the solar time will match the clock and wonder, "How does it do that?" They will, of course, notice that the sundial clock shows an hour difference from mechanical or digital clocks during the daylight saving time period of the year. If you start before the first Sunday in March and send your students out to record the time, the difference will be obvious and dramatic when the time changes. Thus they can see that the Sun has not jumped ahead but it is merely the clocks that have changed. You can ask them if they have noticed any other differences in their lives or in their pets due to the time change. If there are infants in their families, are there any differences in their behaviors? How have the children's free-time activities changed?

Actually, a sundial is fairly accurate if it is placed correctly and has the proper angle of the upright gnomon, based on your latitude. You will need a table from an almanac for your latitude and longitude to make local corrections. If this sounds complicated, it is, but it is fun for older children, though a bit too difficult for younger ones. The digital clocks of the world (like those on our computers hooked up to the internet) now set their time according to a cesium atomic clock, but you probably use a timepiece based on the 24-hour celestial time standard (like our wall clocks and watches that must be corrected occasionally). The atomic clock in the United States is kept in Boulder, Colorado, at the National Institute of Standards and Technology (NIST) but other cesium clocks are located in many parts of the world. You can get one if you have about $20,000 lying around. The creators claim that it loses or gains only one second in 20 million years. As I write this, I notice that my computer clock is right on the mark, but that my digital wristwatch is six seconds fast. But I think that is close enough for me, so I will not bother changing it. Normally we don't care to be exactly accurate but for astronomers and other scientists, it is often important to be absolutely correct. Best to go by the official site at *www.time.gov* to get the correct time within the United States or to the world clock site for times all over the world.

The immense amount of controversy about the benefits of daylight saving time offers a great opportunity for your students to debate the issues. They can obtain a lot of information on the internet or in the library. Writing a story about what it is like to have your clocks changed will allow you to integrate literacy and science lessons. If your students are technically savvy, they can present a Power-Point presentation on their findings. This is also an opportunity for them to conduct a poll among the various grade levels of students in your school and display their findings for the entire school to view using graphs of all sorts, which allows for integration of math and science. Indiana is a very interesting state to research because it has varied time zones within this one state. "What Time is it in Indiana?" is an eighth-grade student-run study, also available on the internet *(www. mccsc.edu/time.html)*.

If you are willing to let your students use electric, heating, or cooling bills as data, you can conduct your own study on the effects of the time change on energy savings. There are data available from your local energy provider that can be used. Crime prevention and auto accident statistics are also available so that local differences can be researched. Interviews are appropriate for human opinion data. Do people feel that they are losing sleep due to the change? Are cows adjusting to the milking time or are farmers just operating by the Sun as usual? These are interesting questions. One jokester even suggested that daylight saving time caused global warming because of the extra hour of sunlight. Even more bizarre was that some people actually believed it. All in all, this story can provide an opportunity for your students to gather a great deal of data and then draw their own conclusions.

In the Teachers Domain through WGBH in Boston there is a wonderful video available to teachers titled "How Clocks Work." You can join this group by going onto the internet at *www.teachersdomain.org/resource/vtl07.math.measure.time. lpclocks*. The video is a cartoon of kids trying to figure out how to keep time when there are no clocks available, to save a friend in danger. I recommend it highly.

related books and NSta journal articles

Driver, R., A. Squires, P. Rushworth, and V. Wood-Robinson. 1994. *Making sense of secondary science: Research into children's ideas.* London and New York: Routledge Falmer.

Keeley, P. 2005. *Science curriculum topic study: Bridging the gap between standards and practice.* Thousand Oaks, CA: Corwin Press.

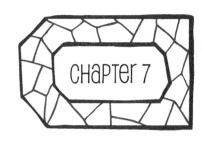

Keeley, P., F. Eberle, and C. Dorsey. 2008. *Uncovering student ideas in science: Another 25 formative assessment probes, volume 3.* Arlington, VA: NSTA Press.

Keeley, P., F. Eberle, and L. Farrin. 2005. *Uncovering student ideas in science: 25 formative assessment probes, volume 1.* Arlington, VA: NSTA Press.

Keeley, P., F. Eberle, and J. Tugel. 2007. *Uncovering student ideas in science: 25 more formative assessment probes, volume 2.* Arlington, VA: NSTA Press.

Konicek-Moran, R. 2009. *More everyday science mysteries.* Arlington, VA: NSTA Press.

references

American Association for the Advancement of Science (AAAS).1993. *Benchmarks for science literacy.* New York: Oxford University Press.

Konicek-Moran, R. 2008. *Everyday science mysteries.* Arlington, VA: NSTA Press.

National Research Council (NRC). 1996. *National science education standards.* Washington, DC: National Academies Press.

WGBH Educational Foundation Teachers Domain. How clocks work. *www.teachersdomain.org/resource/vtl07.math.measure.time.lpclocks.*

CHAPTER 8
SUNRISE, SUNSET

"If you are lost in the woods, and it is not a cloudy morning, how can you tell your directions without a compass?" asked the scoutmaster during the weekly meeting of a Boy Scout troop in northern Idaho.

"Moss only grows on the north side of trees," responded one of the troop members.

"Is that really always true?" probed the scoutmaster.

There was a moment of silent doubt among the 12 wriggling boys.

"That's what I heard, somewhere," came the hesitant reply.

"Well, what if you see lots of trees and not all of them have moss on the same side?"

Silence.

"What about the Sun?" prompted the scoutmaster.

Silence.

"Come on now, you all know where the Sun rises each morning, don't you?

Silence.

Then a weak and unsure response came from the back of the room. "The Sun always rises in the east?"

"And where does it set?"

"In the West … I think," said the young tenderfoot, a first-year scout.

"Right!" said the scoutmaster. "So, if you see the Sun rise or set, you can tell true east or west. That means when we go on our campout this weekend, we'll try to find our way without compasses by using the Sun."

The weekend came. All the boys went out to the campsite, set up their tents and prepared for a hike the next morning. It was a lovely early June morning, not a cloud in the sky, and the Sun came up with a beautiful golden glow on the horizon.

"There it is boys," said the scout leader, "And what direction is that?"

"East?" the boys responded in unison.

"Yep! And now let's just check it out on the compass, so you know that it's true."

Compasses came out of pockets and backpacks and there was a lot of moving around and pointing of the little instruments in all directions. There was also a lot of mumbling.

"Whoops," said one of the boys. "It looks like it is quite a bit north of east to me."

"Let's see how you are reading that compass," said the leader who brought his own compass over next to him and looked in the direction of the rising Sun. Then, he frowned. "Must be something wrong with our compasses."

But, no, each and every scout's compass reported the same result. The Sun was definitely coming up to the north of east, not directly east.

"Well, at least we know that tonight the Sun will set in the west. Even all of the campfire songs tell us that."

That night as the Sun sank beneath the horizon, the boys, a little less sure of their scoutmaster's prediction, checked it out with their compasses just to make sure. What they found gave them a new mystery to solve, one that would take some time and a lot of observations.

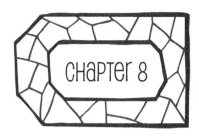

PURPOSE

Astronomical rules are not always correct, especially when they use the word "always." Students should learn that unless they live on or very near the equator, there are only two days in the year that the Sun rises directly in the east—the spring and autumn equinoxes. Another purpose of the story is to help understand the importance of latitude in seasonal solar measurements.

RELATED CONCEPTS

- Equinox
- Solstice
- Latitude
- Earth tilt

DON'T BE SURPRISED

Your students probably have heard that the Sun always rises in the east and sets in the West. Unless they have had the opportunity of mapping the sunrises and sunsets in their area, they are probably unaware that the tilt of the Earth affects the apparent direction of these. They may not know that people in the higher latitudes in the northern hemisphere and the lower latitudes in the southern hemisphere see a different path of the Sun than people at the equator. Finally, your students might not be aware that there is a difference between magnetic north and "true" north.

CONTENT BACKGROUND

It is the tilt of the Earth that causes many of the puzzling patterns that we see of our celestial partners, the Sun and the Moon, as they appear to move about our sky. The Earth does not revolve around the Sun straight up and down but is tilted 23½° from the vertical and remains in the same plane as it makes its revolution around the Sun each 365+ days, our solar year. (To visualize Earth's tilt, think of the standing globes where you see the Earth leaning a bit to the left.) This causes the seasons, although it seems that much of the population is still under the impression that we are closer to the Sun in summer than winter, and that this is the cause of the seasons. If you would like to see more about this misconception, view the *A Private Universe* video on the Annenberg Channel on your computer: *www. learner.org/resources/series28.html.* This fascinating video will allow you to view how in 1987 Harvard graduates—and faculty—had trouble explaining the cause of the changing seasons (Schneps 1987).

We in the northern hemisphere are actually closer to the Sun in the winter; however, the tilt of the Earth's axis allows more *direct* sunlight to the southern hemisphere, which is in the direct line of the Sun's ray. The direct rays of the Sun provide more radiation than indirect rays do, so those areas that receive the more

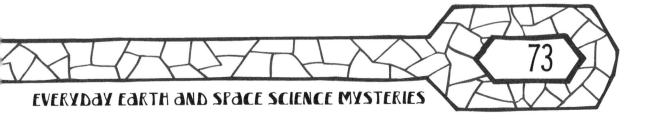

direct rays experience warmer weather. Thus, when the people in the northern United States and Canada have winter, people in Argentina have summer.

But this story is about sunrise and sunset and what we expect of these phenomena with which we are so familiar. Our scoutmaster, and so many others of us, failed to notice the changes that occur every day in those celestial paths. A true observer is one who sees things even though he or she is not seeking to see something in particular. Most of us walk right by many natural phenomena every day and miss them entirely. Our senses are focused on whatever we are doing. Things can happen right before our eyes and we miss them because we are not accustomed to being true observers. One example of this is the Sun and Moon as they make their daily paths across our sky.

If the axis of the Earth were not tilted, every day of the year would be an equinox, with every day having an equal number of lighted and unlighted hours. If we parse the word *equinox,* we find it means "equal nights." Day and night—the number of hours lit and unlit by the Sun—is caused by the *rotation* of the Earth on its axis. But since this axis is tilted in the yearly revolution around the Sun, in the northern hemisphere the vernal equinox marks the beginning of the spring season. The number of daylight hours will begin to increase and nighttime hours will decrease. The land and water in the northern hemisphere begins to point more and more toward the direct rays of the Sun, becoming warmer. The Sun seems to rise higher in the sky as it goes through its daily "path" and the point on the horizon that marks its "sunrise" moves farther north. It will appear to be above the horizon for longer amounts of time as we approach the summer solstice.

Around June 21 in the northern hemisphere, when the axis of the Earth is pointed more directly at the Sun and it is above the horizon for the maximum number of hours for the year, we celebrate the summer solstice, or the first day of summer. In earlier cultures, this day was called *midsummer,* a time of supposedly magical importance. Ancient cultures marked this day, along with those of the equinoxes and the winter solstice, with monuments such as Stonehenge and its counterparts in the Inca empire of south and central America, as well as in the southwest at Chaco Canyon, in what is now New Mexico. Stones were set to mark the passage of the Sun's rays at sunrise so that the rays passed onto or through special passages between stones on particular days. It seems as though the ancient people were much more aware of the seasonal changes than are we today.

It may be advantageous to imagine that you are watching this celestial pageant from some distant place in space. From your vantage point, you can witness the path of the Earth, with its axis always pointing in one direction, as it revolves around the Sun. You can see the two equinoxes when the Sun's rays fall on the Earth so that exactly half of the Earth is lighted. You would notice that at those points, the Sun would rise everywhere on the globe directly in the east. And you would witness the solstices when either the northern or southern hemispheres are favored by more direct light. You would also notice that in the northern hemisphere, the summer solstice would cause the sunrise to appear to be north of east and that the Sun would appear to make a prolonged path across the daily southern sky. At the opposite end of the orbit in the southern hemisphere, you would notice that the sunrise would appear to be south of east and make a prolonged path through the northern sky. If you still have trouble envisioning this, check out the animation at *www.mathisfun.com/earth-orbit.html*.

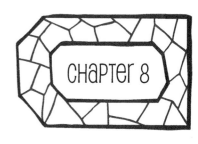

Figure 8.1. Solstices and Equinoxes

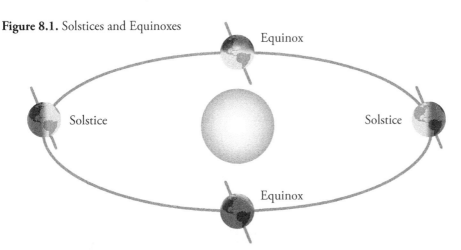

Our scouts were able to see differences between true east and the northeast sunrise spots, especially in the northern latitudes of Idaho. If they had been further north and closer to the Arctic Circle, the difference would have been even greater. Had they been above the Arctic Circle, the Sun would have never set since the axis of the Earth was pointing directly at the Sun as it made a 24-hour circle around the low horizon. By the way, the scoutmaster might have taken the opportunity to explain that he was using a generality and admit his mistake, but we all know how hard that is!

I want to mention the difference between magnetic north as shown on a compass and "true north," which is the direction from your present location to the pole as indicated on a map. In the southern hemisphere, the difference would be between magnetic south and "true south" of the South Pole. The magnetic poles in this giant magnet we call Earth have shifted as much as 1,200 miles over time. Therefore, the compass will not give you the map-related north or south you need. This is why you need to get a topographic map of the area you are interested in from your local sporting goods store or from the U.S. geological survey. They can be found online at *www.usgs.gov*. Topographic maps give not only contour lines but also a reading of what is called *magnetic declination*. Declination is a number of degrees east or west by which you can make corrections between what is shown on your magnetic compass and the true "map" directions. You either add or subtract numbers from the compass reading to head you in the right map direction. For example, you may find the number –10 or W 10 at your location which means that magnetic north is 10 degrees to the west and you need to *add* 10 degrees to your compass reading to find true north. If the reading is +10 degrees or E 10 degrees, the opposite correction must be made to set you on your correct map direction.

In truth, the boys would have had to refer to their maps before using their compasses to find true east. But the point is still that the sunrise would have been toward the northeast and would probably have been very obvious despite the magnetic declination.

Now comes the kicker. I wonder how many of you realize that the Earth's so-called North Pole is really the *south* magnet pole. The pole is named the North Pole merely because it is in the northern hemisphere. The compass needle is a north pole of a magnet and therefore would be repelled by another north pole. It seeks the north because the north magnetic pole is really the south pole of the Earth's magnetic field. However, it makes no difference in how we view our directions. We still head toward the north pole if we are going north and vice versa. And we still say that the Sun in the northern hemisphere rises each day more toward the northeast as the days progress from the equinoxes to the summer solstice. Don't forget those folks who live in the southern hemisphere who are "standing on their heads" and seeing everything 180° differently than we are.

related Ideas From The National Science education Standards (NrC 1996)

K–4: Changes in Earth and Sky

- Objects in the sky have patterns of movement. The observable shape of the moon changes from day to day in a cycle that lasts about a month.

5–8: Earth in the Solar System

- Most objects in the solar system are in regular and predictable motion. Those motions explain such phenomena as the day, the year, phases of the Moon, and eclipses.

related Ideas From Benchmarks For Science Literacy (aaas 1993)

K–2: The Earth

- The Moon looks a little different every day, but looks the same again about every four weeks.

6–8: The Earth

- The Moon's orbit around the Earth once in about 28 days changes what part of the Moon is lighted by the Sun and how much of that part can be seen from the Earth—the phases of the Moon.

USING THE STORY WITH GRADES K-4

The global implications of this story may be too complex for very young children. However, it should raise some interesting questions about the path of the Sun for even the youngest children and give you an opportunity to do some scientific observation and recording. If you read the introduction in this book, you will see how a second-grade teacher used another story, "Where are the Acorns?" to guide her students toward understanding the pattern of the Sun's shadows over a whole school year. Some of the techniques used by the teacher could be applied to this story since many of the concepts involved are similar in both. If you are able to obtain a copy of *Everyday Science Mysteries* (Konicek-Moran 2008), the chapter "Where Are the Acorns?" has a great deal of information about using the story with younger children. I would also suggest that you purchase a copy of *The Old Farmer's Almanac* (Yankee Publishing) for the current year. It is very inexpensive and has tables of astronomical events that can be adjusted for your specific location. It contains accurate tables for sunrise, sunset, moonrise, moonset, solar and lunar eclipses, length of day, and many other useful data for teaching any astronomical topic.

With young children, the first step toward understanding that the Sun travels in a predictable pattern lies in observing shadows and learning about how they can change direction and shape. Have children place a pencil in a vertical position on their desks using a piece of clay to hold the pencil in place. Give each team of children a flashlight and try to dim the lights in the classroom. Let them experiment with the flashlight and the shadows created by the pencil when they move the light source. Ask them to do the following:

- Make the shadow move toward your left hand.
- Make the shadow move toward your right hand.
- Make the shadow get longer.
- Make the shadow get shorter and shorter until it disappears.

Have students record or draw where their light source was in each case. Then ask them to help you create a list of things they have learned about shadows. Post these on chart paper and if there are disagreements, test them to settle the discrepancies.

At this point, they are usually ready to go outside and observe their own shadows at different times of the day. They can also play "shadow tag" where they must tag another player's shadow with their own. All of these activities will make them more familiar with how shadows are formed and how they change.

Moving to shadows made by the Sun is the next step. They should now place a stick on level ground so that it is vertical. This is known as a *gnomon* (Figure 8.2). The word comes from the ancient Greek and means indicator. It is the part of the sundial that casts the shadow. Once they see that the Sun is their light source, the motion and length of the shadows will correspond to their activities with the flashlight.

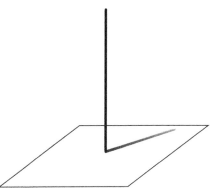

Figure 8.2. Gnomon

The big difference is that they cannot control the Sun like they did the flashlight, so they must wait for the Sun to move and change the shadow. This means that they must record the shadow position and length on a piece of paper. The easiest way to do this is to stick the gnomon through the middle of a sheet of paper and secure the edges of the paper with rocks or toothpicks. At various intervals during the day, children can record the time and draw over the shadow to record their lengths and directions.

As children get used to the apparatus, and perhaps for children who are a bit older, they can see that in the northern hemisphere, after the equinox in March, the morning shadows will retreat a bit more toward the south each day in comparison to where they were earlier in the month at the same time. This can be translated to mean that the Sun is farther north in the sky so that the opposing shadow appears farther south on the paper. Of course, in the southern hemisphere the directions would be reversed.

They may also discover that as the summer months approach, the shortest shadow of the day becomes shorter, meaning that the Sun is becoming higher in the sky at midday. If you should happen to teach on Key West in southern Florida, you are only one degree north of the Tropic of Cancer which is 23.5° north of the equator. On June 21(the summer solstice) the Sun will be very close to being directly overhead at midday on that date. But that is as far north as the overhead Sun will get. In other words, the myth that the Sun is directly overhead everywhere at noon is false.

Reasons for the seasons should probably be reserved for the middle school years, and we will continue suggestions for this in the next section.

USING THE STORIES WITH GRADES 5–8

My experience with students in grades 5–8 is that some, but not all, are able to use the spatial relationships necessary to understand such things as the reasons for the seasons and the phases of the moon. I do believe that it is worth a try but do not be surprised if they finally "get it" sometime later in someone else's class. You can be satisfied that you did your best and that whatever you did laid the groundwork for the final "Aha!" that is bound to come later.

I believe it would be helpful to you to view the *A Private Universe* video (Schneps 1987) mentioned previously and see it for yourself before you begin to work with your students. You will see students, not much older than your own, struggle with the ideas of the interactions of the Sun, Moon, and Earth. You'll notice that one of the students quickly used props to help her understand how the three celestial globes related to one another. Although not all students have a tactile learning style, it usually helps most students to play with objects and to maneuver them to help them to build a visual model of the positions that cause things like the seasons, moon phases, and eclipses. Even though this is not the

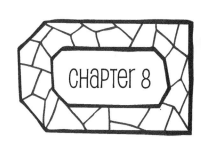

main purpose of this story, it helps if the students have some understanding of how the Sun's rays, as they impact the Earth, change over a period of a year.

If you would like to introduce your students to orienteering or using maps to find their way, you will need to teach them compass skills and make them aware of the magnetic declination issue described previously. Sometimes, orienteering is a great way to engage students in discovering the intricacies of the magnet upon which we spend our lives.

Students in middle school should now be able to finish the story about the scout trip and make sense of the problem caused by the discrepancy between the scoutmaster's general statement and what they found with their compasses. They may want to do some checking on the equinox to see if it is true as well that the sun rises directly in the East. After all, checking patterns for oneself is a major part of science. Also point out the importance that it is a June camping trip in the story (well past the solstice).

related Books and NSTa Journal articles

Keeley, P. 2005. *Science curriculum topic study: Bridging the gap between standards and practice.* Thousand Oaks, CA: Corwin Press.

Keeley, P., F. Eberle, and C. Dorsey. 2008. *Uncovering student ideas in science, volume 3: Another 25 formative assessment probes.* Arlington, VA: NSTA Press.

Keeley, P., F. Eberle, and L. Farrin. 2005. *Uncovering student ideas in science, volume 1: 25 formative assessment probes.* Arlington, VA: NSTA Press.

Keeley, P., F. Eberle, and J. Tugel. 2007. *Uncovering student ideas in science, volume 2: 25 more formative assessment probes.* Arlington, VA: NSTA Press.

Keeley, P., and J. Tugel. 2009. *Uncovering student ideas in science, volume 4: 25 new formative assessment probes.* Arlington, VA: NSTA Press.

Konicek-Moran, R. 2009. *More everyday science mysteries: Stories for inquiry-based science teaching.* Arlington, VA: NSTA Press.

Konicek-Moran, R. 2010. *Even more everyday science mysteries: Stories for inquiry-based science teaching.* Arlington, VA: NSTA Press.

references

American Association for the Advancement of Science (AAAS).1993. *Benchmarks for science literacy.* New York: Oxford University Press.

Driver, R., A. Squires, P. Rushworth, and V. Wood-Robinson. 1994. *Making sense of secondary science: Research into children's ideas.* London and New York: Routledge Falmer.

Gilbert, S. W., and S. W. Ireton. 2003. *Understanding models in earth and space science.* Arlington, VA: NSTA Press.

Keeley, P., and J. Tugel. 2009. *Uncovering student ideas in science, volume 4: 25 new formative assessment probes.* Arlington, VA: NSTA Press.

Konicek-Moran, R. 2008. Where are the acorns? In *Everyday science mysteries: Stories for inquiry-based science teaching,* 39–50. Arlington, VA: NSTA Press.

Konicek-Moran, R. 2009. *More everyday science mysteries: Stories for inquiry-based science teaching.* Arlington, VA: NSTA Press.

National Research Council (NRC). 1996. *National science education standards.* Washington, DC: National Academies Press.

Oates-Brockenstedt, C., and M. Oates. 2008. *Earth science success: 50 lesson plans for grades 6–9.* Arlington, VA: NSTA Press.

Schneps, M. 1986. *A private universe project.* Harvard Smithsonian Center for Astrophysics.

United States Geological Survey. *www.usgs.gov*

Yankee Publishing. *The old farmer's almanac,* published yearly since 1792. Dublin NH: Yankee Publishing.

NATIONAL SCIENCE TEACHERS ASSOCIATION

CHAPTER 9
NOW JUST WAIT A MINUTE!

The science challenge had been set and the whole school was excited about competing. This year, instead of the usual science fair, the science committee asked each student competing to meet a special challenge in any way they could think of. They could work in teams or as individuals, but they had to follow strict rules, including how much they could spend and what materials they could use.

The challenge was to build a timing device that would measure exactly one minute. It could not have any clock mechanisms in it and had to rely strictly on a force to make it work—any force the participants chose. It could not cost more than $5 in materials. Judging would be based on how close each team's timer came to measuring exactly one minute!

Enrique and Katerina wanted to work together but they had very different ideas on how to make the timer and what forces to use. Enrique wanted to use gravity and Katerina wanted to use buoyancy, the force that makes certain things float or sink.

"I want to make a ramp for a marble that takes one minute to roll down a maze," said Enrique. "It will use the force of gravity."

"That's too easy," said Katerina. "We ought to use a water clock. That's a bowl that has a hole in the bottom of it and floats in another bowl for only one minute and then sinks."

"I know what a water clock is but that's pretty easy too," said Enrique, "So maybe we can figure out something that is really cool and something nobody else would think of."

"We ought to be able to come up with something new and different. Anyway, if we can't think of anything, we can always fall back on the water clock."

"Or the ramp," added Enrique.

"Okay, or the ramp," replied Katerina without much excitement in her voice. "One way or the other we ought to be able to make something that will do its thing in just one minute. How many forces are there that we can use to make our timer work? And how do we get it to do whatever it does in exactly one minute?"

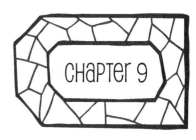

PUrPOSE

This story obviously is aimed at the technology standards. Two simple timing devices are mentioned with the suggestion that more are possible. These can be improved to meet the challenge or other devices could be invented. Students are being challenged to either improve on an idea that already exists or to be creative and think of another type of timing device. To judge this challenge, it is a good idea to have a stopwatch that is accurate to at least tenths of a second. I have witnessed contests that were decided by just this interval! Students often become very creative—and competitive—in meeting this challenge.

reLaTeD CONCePTS

- Time
- Forces
- Accuracy
- Design

DON'T Be SurPriseD

Time, especially for young children, is merely something that passes between important events. Long time intervals may be between holidays, while short time intervals may be between recess and lunchtime. How many parents have heard the question "How much longer?" when traveling or waiting to go somewhere? Clocks may have a meaning for some young children but even more effective are the many devices available based on the hourglass design. Somehow, seeing sand pass through a hole from top to bottom gives them a sense of beginning and end. Many of them will have used these devices in games they have at home or school.

Time often seems something set by adults. Many children cannot fathom why they can't see their favorite TV show when they want to see it rather than at the scheduled time they feel to be arbitrary. Also, don't expect your young students to connect the astronomical time clock to the daily running of their lives. Setting the clocks forward or backward for daylight saving time rarely changes the meaning of time as far as they are concerned. This is even so for some adults who still believe that the "extra hour of daylight will not be good for the crops," (a quote from a friend of my grandfather, years ago).

CONTeNT BaCKGrOUND

The measurement of time may have begun at the dawn of our species' arrival on this planet. Certainly, early humans were aware of the distinction between night and day and probably worshipped gods whom they believed controlled these happenings. The patterns of the celestial bodies that were so important to their lives did not go unnoticed or unrecorded. We have evidence of the significance of

such events as solstices and equinoxes in structures that still survive. Stonehenge in southern England was a celestial clock and perhaps a place of worship. The stones in Stonehenge are arranged to coincide with various celestial happenings, including the equinoxes and solstices. Early Anasazi engineers in Chaco Canyon in northern New Mexico also developed structures for marking celestial happenings over a thousand years ago. And of course the Aztecs and Mayans of South and Central America had celestial calendars as early as the 11th century, CE (common era). These calendars were remarkably accurate and predicted important holidays, festivals, and of course the significant agricultural events of the year.

It is easy to infer from existing structures that from the earliest civilizations on, time was connected to the periodic motion of celestial bodies such as the Sun, the Moon, and the various constellations that were visible at different times of the year. Early Egyptians used obelisks as sundials and even developed elaborate water clocks called *clepsydras* (water thieves), which marked the day of important periods. Almost 2,000 years ago, in the Chinese empire, water clocks of great complexity were built and used to measure the passing of time, both day and night.

When clocks were built, they had to have certain properties, the most important of which was the *consistent, periodic, constant,* and *repetitive* action that could mark off equal increments of time. Due to issues such as pressure and changes in dynamics, this was no easy matter and many designs were attempted over the centuries. For example, as water from a container dripped into another container, the pressure of the water could change causing the water to drip more slowly as time went on. Our ancestors were creative and clever and invented many ways of measuring time as the need for more accuracy evolved.

In more recent times, during the 1760s, an Englishman named John Harrison developed a clock and watch that kept such excellent time that it could be used aboard ships to ascertain longitude for oceangoing travel. His watch was accurate to five seconds over a period of months. And as a result, the pocket watch was born and became the standard of timekeeping for hundreds of years, even among the poorer citizens of the world.

The history of how various civilizations have used and manipulated time is a fascinating story and one worth learning. Travel and commerce dictated the need for accurate timetables, time zones, and all of the time-related manifestations of rules and regulations. The science of astronomy and the pseudoscience of astrology, politics, and religion have all contributed to time measurement and manipulation over the centuries.

Now in the 21st century, the American government keeps the standard of time with a cesium atomic clock that is so accurate that it will not lose or gain a second in 60 million years! (What's that old joke about "close enough for government work"?) The clock uses the regular movement of cesium atoms to calculate the time intervals. It's much more complicated than that, but this is a short book! You can find that exact time on the internet at *http://nist.time.gov*, in case you want to set your watch. As you can see, we have come a long way from obelisks and clepsydras and circles of stones. But if you expect that time scientists and engineers

are satisfied with an accuracy of one second in 60 million years, you have another thing coming. They are constantly trying to improve their invention, just as people have been trying to improve timekeeping inventions since the dawn of civilization. We are therefore asking your students to participate in the fun of improvement and to join those scientists of the past who tried to develop timekeepers that kept accurate time.

Two timekeepers are mentioned in the story, the maze and the water clock. The maze is merely a platform that allows gravity to pull a marble down through a maze of baffles which slow it down so that it reaches the bottom at a prescribed time rather than just plummeting down in fractions of a second. The baffles are modified by trial and error so that the marble takes the desired amount of time to reach the bottom.

Water clocks can be of several designs but essentially work on the basis of dripping water or water entering through a hole at a constant rate. One clock is a bowl with a small hole in the bottom which is set in another bowl of water. The hole allows water to enter and eventually sinks the bowl. The rate of sinking is controlled by the size of the hole. Another modification is a bowl that drips water into another bowl until it sinks the second bowl. Again the rate is determined by the size of the hole.

There are timekeepers that are made from candles that burn at a reliably constant rate. I have seen candles placed on a seesaw-like balance so that the balance moves as the candle burns and changes the balance between the two ends.

I return to the common properties of the timekeeper mentioned before, a consistent, periodic, constant, and repetitive action that will measure the intervals of time required. These properties can be interpreted liberally, such as in the case of the maze. But the main point is that the timekeeper must work consistently and not change over time. Your students should be able to come up with a myriad of designs to compete with others for the most accurate timekeeper.

related ideas from the National Science Education Standards (NRC 1996)

K–4: Abilities of Technological Design

- Identify a simple problem.
- In problem identification, children should develop the ability to explain a problem in their own words and identify a specific task and solution related to the problem.
- Propose a solution.
- Students should make proposals to build something or get something to work better: They should be able to describe and communicate their ideas. Students should recognize that designing a solution might have constraints, such as cost, materials, time, space, or safety.

5–8: Abilities of Technological Design

- Design a solution or product.
- Students should make and compare different proposals in the light of the criteria they have selected. They must consider constraints—such as time, trade-offs and materials needed—and communicate ideas with drawings and simple models.
- Implement a proposed design.
- Students should organize materials and other resources, plan their work, make good use of group collaboration where appropriate, choose suitable tools and techniques, and work with appropriate measurement methods to ensure adequate accuracy.
- Evaluate completed technological designs or products.
- Students should use criteria relevant to the original purpose or need, consider a variety of factors that might affect acceptability and suitability for intended users and beneficiaries, and develop measures of quality with respect to such criteria and factors; they should also suggest improvement and, for their own products, try proposal modification.

related ideas from Benchmarks for science Literacy (aaas 1993)

K–2: Technology and Science

- Tools are used to do things better or more easily and to do some things that could not otherwise be done at all. In technology, tools are used to observe, measure, and make things.
- When trying to build something or to get something to work better, it usually helps to follow directions if there are any or to ask someone who has done it before for suggestions.
- People alone or in groups are always inventing new ways to solve problems and get work done. The tools and ways of doing things that people have invented affect all aspects of life.

K–2: Designs and Systems

- People can use objects and ways of doing things to solve problems.

3–5: Designs and Systems

- Even a good design may fail. Sometimes steps can be taken ahead of time to reduce the likelihood of failure, but it cannot be entirely eliminated.

3–5: Technology and Science

- Throughout all of history, people everywhere have invented and used tools. Most tools of today are different from those of the past but many are modifications of very ancient tools.
- Any invention is likely to lead to other inventions. Once an invention exists, people are likely to think up ways of using it that were never imagined at first.

6–8: Designs and Systems

- Design usually requires taking constraints into account. Some constraints such as gravity or the properties of the materials to be used are unavoidable.
- Technology cannot always provide successful solutions for problems or fulfill every human need.

USING THE STORY WITH GRADES K–4

You may have to remind the students of ways they keep time when playing games. It would be very surprising if some, if not all, of the students did not have at least a little experience with sand timers. Depending upon their motor skills level, you might ask them if they can make something similar themselves using pill bottles or other containers and sand. Ask them how they might make the timer keep longer time intervals or shorter time intervals. They usually suggest bigger containers, or more or less sand, or sometimes a change in the size of the hole through which the sand flows.

Younger children probably do not have the small motor coordination necessary to build a maze and marble ramp. However, the toy ramp and marble games that some children possess might be modified to take different times to operate. One is available from Sense Toys with a ramp that can be changed for slower or faster action. Lego blocks can also be used to create ramps which can be modified by the children. Another idea would be to use car ramps, which usually come with lots of extensions and curves and can be adjusted easily. Children may think that heavier balls will take longer to roll down the ramp and this can give you an opportunity to test this hypothesis.

My experiences with fourth graders and older have been very successful. They often want to make the maze out of wood but finally realize that box cardboard will work just as well and keep the cost down. They realize that the tilt of the maze board has a great effect on the timing device. Some have suggested placing a metal cup at the end so the ball will drop, signaling the end of the run. They also enjoy designing a water clock and soon realize that modifying the hole in the bottom of the bowl with tape has an effect on the length of time the bowl stays afloat. Some with more creative and/or technological bents will try to make tipping devices and/or candles with markings. The latter may prove difficult to read and are usually abandoned.

One of the most ingenious devices I have seen involved a candle on one end of a balance that tipped as the candle burned and lost weight. The balance was placed next to a scale of times marked on a piece of paper. It worked for several trials, but then it toppled over—a little engineering drawback! You may well be amazed at the different ideas that emerge from your students.

Be sure to take this opportunity for the students to use their science notebooks to record the challenge as they see it, their plan, and of course their results and conclusions. A list of problems incurred during the testing should also be included along with a narrative of how they went about trying to solve each setback. Their problem-solving techniques often provide the best discussions about techniques used in engineering analysis and action. Inviting an engineer into the classroom to talk with the children can also be a rewarding experience, both for the class and the engineer. Some engineering companies belong to nationwide organizations that help classroom teachers promote engineering curricula. Lockheed-Martin, Cisco, and Intel are just some of the sponsors of such enterprises. Look them up on the internet to find out how they can help you in advancing your technological curriculum.

In the March 2007 issue of *Science and Children* there are two trade books recommended by Christine Anne Royce and ideas for using these books for K–3 children. The books are: *Let's Try It Out With Towers and Bridges* and *Bridges: Amazing Structures to Design, Build, and Test*. I heartily recommend your looking into this article and the following suggestions for use with K–3 children. Although it does not directly apply to the story, it may offer a nice introduction into children building new structures to apply to problems that need to be solved.

USING THE STORY WITH GRADES 5–8

First, I would suggest that you read the K–4 section including the part about engineers in the classroom. Having worked with these organizations, I find them to be sincere and very skillful in working with students and teachers.

Middle school students usually jump at a chance to meet a challenge such as the one described in the story. You and the students can make up a rubric for assessing the success of each project. This rubric would allow some leeway for ideas that did not win the challenge but were within an agreed-upon range of accuracy and used good problem-solving techniques. There could also be a criterion for creativity and for the most unusual creation. Many of the devices mentioned above can be created as well as the oft-used "domino-effect timer" made by stacked dominos knocking each other over for a given amount of time.

Another possible diversion can be a Rube Goldberg (1883–1970) contest, named for the cartoonist of the mid-20th century who was famous for creating depictions of complex devices for completing simple tasks. An excellent example

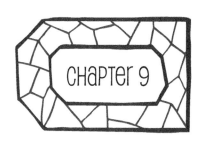

of a Rube Goldberg–type of device is in the famous Honda commercial that can still be found on YouTube. Encourage your students to use their search engines to see videos of Rube Goldberg devices and stimulate their creativity to use a combination of forces to accomplish a task. A book of his cartoons called: *Rube Goldberg: Inventions!* by M. F. Wolfe (2000) is a great resource for the students. The website *rubegoldberg.com* is also the site for national contests and more information about his work. You can even find him in the dictionary as a noun. Look it up and see! As for the educational value of these forays into what may seem like fantasy, the planning and fine tuning of these devices are examples of what engineers have to do to accomplish their jobs while adding the whimsical and fun aspects to the challenge. It also allows students to use their knowledge of forces in unusual ways.

related books and NSTA journal articles

Driver, R., A. Squires, P. Rushworth, and V. Wood-Robinson. 1994. *Making sense of secondary science: Research into children's ideas.* London and New York: Routledge-Falmer.

Keeley, P. 2005. *Science curriculum topic study: Bridging the gap between standards and practice.* Thousand Oaks, CA: Corwin Press.

Keeley, P., F. Eberle, and C. Dorsey. 2008. *Uncovering student ideas in science: Another 25 formative assessment probes*, volume 3. Arlington, VA: NSTA Press.

Keeley, P., F. Eberle, and L. Farrin. 2005. *Uncovering student ideas in science: 25 formative assessment probes*, volume 1. Arlington, VA: NSTA Press.

Keeley, P., F. Eberle, and J. Tugel. 2007. *Uncovering student ideas in science: 25 more formative assessment probes*, volume 2. Arlington, VA: NSTA Press.

Royce, C. A. 2007. If you build it. *Science and Children* 44 (7): 14–15.

references

American Association for the Advancement of Science (AAAS). 1993. *Benchmarks for science literacy:* New York: Oxford University Press.

Johmann, C. A., and E. Rieth. 1999. *Bridges: Amazing structures to design, build, and test.* Charlotte, VT: Williamson Publishing Co.

National Research Council (NRC). 1996. *National science education standards.* Washington, DC: National Academies Press.

Royce, C. A. 2007. If you build it. *Science and Children* 44 (7): 14–15.

Simon, S., and N. Fauteux. 2003. *Let's try it out with towers and bridges.* New York: Simon and Schuster.

Wolfe, M. F. 2000. *Rube Goldberg: Inventions!* New York: Simon and Shuster.

CHAPTER 10

WHAT'S HIDING IN THE WOODPILE?

The *Rug Rats* cartoon was just getting started as Maddie and Justin were settling in for a cozy morning. The room was warm and bright. The wood stove gave off a heat you could almost see. They loved sitting on cushions here and watching TV on Saturday mornings.

Suddenly they were interrupted by Mom, who came in to put more wood in the stove.

"Uh oh," she exclaimed! "Somebody's been forgetting their chores. The wood box is empty and it needs to be filled right away. That is if folks want to enjoy the nice warm room."

The children pretended to be very involved in the show. No one spoke.

"Ahem," said Mom. "Ahem!" she said a little louder. "Wood? We need wood, now!"

Justin looked up. He sat right next to the huge empty wood box and knew what was going to happen. They were going to have to trudge outside into the frosty morning, wheel loads of wood into the garage, and carry it from the garage into the family room. Then it needed to be stacked in the wood box. From the look on Mom's face, it was going to happen soon.

"You know, if you carried a little bit in each day," Mom explained for about the hundredth time, "it wouldn't be such a big chore. But you let it get empty and now you have to do it all at once. Chores first, then TV."

Maddie and Justin slowly got up from the floor and went to dress for their job. They knew she was right, but somehow, something always came up each day that was more important than bringing in wood. They walked out into what seemed to be the North Pole, shivering as the wind blew past them and swirled snow at their feet.

C'mon," said Justin. "Let's get this over with. We have a huge box to fill."

An hour later, two very rosy-cheeked children put the last log into the box. They were tired but the exercise had made them feel warm.

"There, that's done for a while," said Justin as he closed the wood box lid. "From now on, I'm carrying in my three logs every day. Filling the whole box is more than a chore!"

The children sat down after removing their coats, hats, and mittens. Now for a cozy hour with the cartoons! Maddie sat on her cushion near the TV and Justin on his, near the wood box. After about five minutes, Justin began to shiver.

"Hey!" he said. "It's cold down here!"

"You'll warm up," said Maddie. "You've been outside too long."

Another five minutes later Justin complained again.

"It's colder than ever!" he claimed.

Maddie scooted over to where he sat. She could feel the difference almost at once.

"It's beginning to get cold over by my spot too," she said. "Did you leave a door open?"

Justin checked. All the doors were closed.

"Well, where's that draft coming from?" asked Maddie.

The area around the wood box was definitely chilly. Too chilly for comfort. The call for lunch sounded from upstairs and the two trudged up for their midday meal. Later that afternoon, they came downstairs to a lovely warm room.

"I wonder what made it so cold this morning," Justin pondered. "Now it's fine again."

"Maybe it was the wood," Maddie replied. "But how could wood make it chilly? Wood is supposed to make things warm."

"Yeah, when it's burning," explained Justin. "But what about when it's just sitting there?"

"Well," said Maddie, "it's just sitting now, and it's not cold in here anymore. What's the difference?"

PURPOSE

Wood comes from trees, right? It is full of potential chemical energy that can produce heat when it is put into a stove or fireplace and burned. Yet, in this story, the Earth's bounty seems to produce a cooler room. What can be the cause? Thermodynamics, the branch of physics that deals with the conversion of various forms of energy from one to another, affecting things such as temperature, is the answer to this riddle, since the flow of heat goes from warmer to cooler objects in a closed system. Once again, this story parallels personal experience, since my children and I also brought in a large quantity of wood from the cold outdoors into a warm room. This story can lead to a great discussion of our uses of the words *warm* and *cold* in the colloquial sense as compared to the words physicists use. It can raise questions about the wood created with carbon from the atmosphere and the Sun's energy and its potential to "produce" heat or, in this case, to absorb heat. It can also provoke inquiry about the nature of heat, temperature, and how heat moves from one substance to another.

related Concepts

- Thermodynamics
- Temperature
- Energy transfer
- Specific heat
- Energy
- Cooling and heating
- Heat

DON'T BE SURPRISED

Children have many misconceptions about heat and temperature. They believe that the two concepts are the same thing. If they believe that wool mittens contain heat, why not the same misconception about wood? (See the article my colleague, Bruce Watson, and I wrote about this in the *Phi Delta Kappan*, 1990. This article can be found online at *www.exploratorium.edu/ifi/resources/workshops/teachingfor-concept.html*.) Although children's experience with wood is mainly with its burning, wood is still seen as a source of heat and not a substance that will cool off a room. Yet a mass of wood possessing a very low temperature is a *heat sink* and absorbs heat energy just like any other cool mass in a warmer environment. Again, our use of colloquial language will tend to cause the children to talk in terms of "cold" as an entity rather than as a *lack* of heat energy. Experience has shown that a lively discussion will ensue after the story is read to the class.

CONTENT BackGround

Our two happy children sitting in the cozy room were unaware of the reason for the warmth in their surroundings. The wood stove with logs alight was *radiating*

heat out into the cooler room and as far as the children were concerned, the room was warm enough for their comfort. *Radiation* occurred when the burning logs heated the metal stove, which gave off infrared rays that traveled both directly to the children close enough to the stove to receive them and into the air molecules in the room. The space between these air molecules began to expand as the molecules got warmer and rose toward the ceiling. These warm molecules were replaced by cooled air molecules tumbling down, which were in turn warmed by the radiation from the stove. Thus a circular current of air moved throughout the room, from floor to ceiling and back again, creating fairly even temperatures for our children in what is called a *convection cell.*

Then things changed as the children brought in a large mass of wood that had lost a great deal of heat in the freezing outdoors. It had become as cold as the air outside the house. One part of the law of thermodynamics states simply that *heat* energy flows from hot to cold. Even though the wood stove was doing its best to keep the room temperature constant, much of the warm air in the room moved toward the cold wood in sufficient amounts to cause a drop in the temperature and even cause what Maddie described as a draft of cold air. Finally, the room warmed up as the temperature difference between the once ice cold wood and the room air reached equilibrium, and since there was no difference between the temperatures of the two, heat only flowed enough to maintain this equilibrium. This flow of heat goes on continuously in any space because everything in that space must attain or maintain an equal temperature. In fact, *heat* is defined as the flow of energy from an object that is warm to one that is cooler.

Before discussing this with your class, I suggest that you read up on the subject of energy transfer in a physical science book. One good option is *Science Matters* by Hazen and Trefil (1991), specifically chapter 2 on energy. Its simple and down-to-earth explanations will give you confidence in your understanding of the concepts used here.

Basically, there are three terms that often are used incorrectly in everyday speech: *thermal energy, temperature,* and *heat. Thermal energy* refers to the total amount of *kinetic energy* in a substance caused by the motion and collisions of the molecules that make up that substance. *Temperature* is a term that refers to the average kinetic energy in a substance and is measured by an instrument called a thermometer, which uses any of the three arbitrary scales: Fahrenheit, Celsius, and Kelvin (named after the people who devised them). Each has its own values and uses but that is another story. And, as we said before, *heat* is the flow of energy from a warmer object to a cooler object.

Let's talk a moment about temperature and thermal energy. Consider a pot of boiling water. When you want to take the temperature of the water, you can put the thermometer only in one place. It will register only the thermal energy of the molecules that strike it at that place. You can assume that because of convection currents, the kinetic energy throughout the pot is fairly consistent. Therefore you would get essentially the same reading wherever you put the thermometer. This is the water's *temperature.*

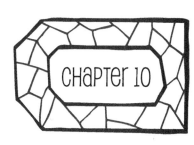

But this is not a measure of the total amount of *thermal energy* in the pot of water. It is only a snapshot of the average kinetic energy of the molecules striking the thermometer at that particular place in the pot. A single drop of that water or the entire pot of water would have the same temperature, but pouring the entire pot over your hand and putting a single drop on your hand would result in very dissimilar outcomes. In other words, the amounts of *thermal energy* transferred to your hand would be seriously different. Even though this analogy is not one we would wish to demonstrate, it does point to the difference between temperature and thermal energy.

Back in the warm room and the addition of the ice-cold wood mass, where we can see that the *thermal energy* that made the room cozy for the children did not stand a chance against the effects of thermodynamics. The *temperature* difference between the huge mass of wood and the air surrounding it dictated the transfer of *heat* from the room air to the wood until they were the same temperature. When that equilibrium is reached the only flow of heat that occurred was that which was needed to reach equilibrium, and that would be more or less constant but not evident to our limited senses. Eventually, everything in the room, with the exception of the stove and the people, who are producing heat, would maintain the same temperature.

So we can see that any cold mass (regardless of the material it is made of) introduced into a warmer environment would produce the cooling effect felt by Justin and Maddie. Some materials do take in thermal energy and give it off at different rates. For example, water requires more energy and time to raise its temperature than many other substances and by the same token gives off that amount of energy more slowly. Metals, however, react in a different manner. It takes less energy to raise the temperature of most metals. They absorb heat faster and release that heat faster than water—or wood, for that matter. Thus, if you were to touch a log and a metal object which had the same temperature, you would find that the metal object felt cooler since it would take heat from your hand faster than the log and thus feel cooler. You may have noticed also that if you are heating water on a stove in a metal pan the pan gets too hot to touch before the water does.

related Ideas From The National Science education Standards (NRC 1996)

K–4: *Properties of Objects and Materials*
- Objects have many observable properties, including size, weight, shape, color, temperature and the ability to react with other substances. Those properties can be measured using tools, such as rulers, balances, and thermometers.
- Objects are made of one or more materials, such as paper, wood, and metal.

- Objects can be described by the properties of the materials from which they are made, and those properties can be used to separate or sort a group of objects or materials.

K–4: Light, Heat, Electricity, and Magnetism
- Heat can be produced in many ways such as burning, rubbing and mixing one substance with another. Heat can move from one object to another.

5–8: Transfer of Energy
- Energy is a property of many substances and is associated with heat, light, electricity, mechanical motion, sound, nuclear energy, and nature of a chemical change. Energy is transferred in many ways.
- Heat moves in predictable ways, flowing from warmer objects to cooler ones until both reach the same temperature.

related ideas from Benchmarks for science literacy (aaas 1993)

K–2: Energy Transformations
- The sun warms the land, air, and water.

3–5: Energy Transformations
- When warmer things are put with cooler ones, the warm ones lose heat and the cool ones gain it until they are all the same temperature. A warmer object can warm a cooler one by contact or at a distance.
- Some materials conduct heat much better than others. Poor conductors can reduce heat loss.

6–8: Energy Transformations
- Heat can be transferred through materials by the collision of atoms or across space by radiation. If the material is fluid, currents will be set up in it that aid the transfer of heat.
- Energy appears in different forms. Heat energy is in the disorderly motion of molecules.

USING THE STORY WITH GRADES K–4

You may be wondering why this story is placed in the Earth and space section of the book when the concepts seem to be more applicable to the physical sciences. My opinion is that the Earth sciences are so directly related to all other sciences

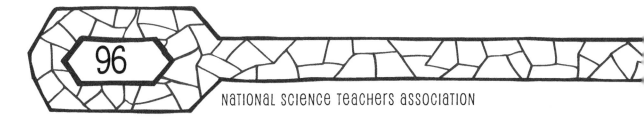

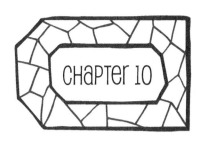

that any discussion of energy in any form is applicable to them. This story deals with energy within the atmosphere and its concepts would fit into looking at energy transformations and transfer, a big conceptual aspect of Earth sciences.

Particularly in the lower grades, where the concept of energy is not specifically addressed by the Standards or the Benchmarks, teaching about energy may present a problem. However, every young child has experienced the effects of higher or lower temperatures whether at home or away from home. Even at the early ages, children can be made aware of the existence of heat and know the difference between heat moving out of something and thereby lowering its temperature and "cold" moving in with the same result.

Helping K–2 students become familiar with a thermometer as a tool for measuring changes in material is a worthy goal. Keeping track of outdoor and indoor temperatures and keeping a weather log is also a wonderful way to connect science and communication skills. Starting early on this sort of learning is important. Students should have the opportunity to connect images and text; observation can be recorded with either or both. Students can plot differences between temperatures by drawing thermometers even if they cannot read and understand the numbers. I have had success in using Centicubes to represent changes in temperature so that children can compare by placing one column next to another. This is, of course, beginning with a qualitative approach to comparison but for very young children it is often the best way.

With grades 3 and 4 there is usually no problem in using a thermometer and they can easily recreate the situation in the story. I suggest using a Styrofoam container large enough to place two to three logs that have been cooled by placing them in a freezer or, if possible, outdoors, if temperatures drop low enough in your area. An indoor/outdoor thermometer is a good way to measure temperatures, with the outdoor probe placed inside the container so that room temperature (which is where the temperature will start before the wood is added) and changes inside the container can be recorded. Predictions are in order here, and hypotheses can be placed on an "Our Best Thinking" chart or in students' science notebooks and kept for future reference.

If you use the method of recording the statements of the children about what they think they understand about the problem in the story on the "Our Best Thinking" chart, you can get statements that can be changed to questions. Questions may include

- Will the temperature change inside the box after the wood is added?
- How long will it take before the temperature stops changing?
- How will fewer or more logs affect the time and temperatures?
- What will happen to the temperature if the lid is removed?
- If the lid is removed, how long will it take before the temperature stops changing?
- What differences, if any, will we find if we use other materials?
- Does the amount of material make a difference in what happens?

Other questions may arise as the activities proceed so be prepared to modify the lesson as the questions dictate. This is, after all, what inquiry is all about.

USING THE STORY WITH GRADES 5-8

Students will probably suggest creating a simulation of the circumstances represented by Justin and Maddie for testing the accuracy of the story. You may suggest the use of the Styrofoam container if they do not. It will probably bring up the idea of insulation and insulating materials. The room the children were in was probably insulated from the outside since that is the building code in most states and is a way of saving on heat bills as well. This brings up an opportunity to talk about why we insulate our homes against heat entering or leaving our living quarters.

I would suggest using the statements of the children about what they predict will happen on the "Our Best Thinking" chart and saving the chart and changing the statements to questions. The questions will probably be similar to those listed above in the grades 3–4 section, but there may be more involving other materials and keeping the variables the same and controlling for differences. The discussions about experimental design will probably be heated since it will be difficult to find materials that are exactly the same mass to test against each other. They should test their hypotheses about the amount of temperature differences expected due to differences in mass.

Again, the inquiry involved here is the goal. Planning and carrying out investigations, which involves interpreting data and drawing conclusions, gives the students a chance to engage in real science. Remember, the story is a stimulus to bring out the students' misconceptions about heat and temperature and allow them to own their investigations. They should have no trouble finishing the story of Maddie and Justin after exploring the many facets of thermodynamics in the atmosphere of the container.

RELATED BOOKS AND NSTA JOURNAL ARTICLES

Damonte, K. 2005. Heating up and cooling down. *Science and Children* 42 (8): 47–48.

Driver, R, A. Squires, P. Rushworth, and V. Wood-Robinson. 1994. *Making sense of secondary science: Research into children's ideas.* London and New York: Routledge-Falmer.

Keeley, P. 2005. *Science curriculum topic study: Bridging the gap between standards and practice.* Thousand Oaks, CA: Corwin Press.

Keeley, P., F. Eberle, and C. Dorsey. 2008. *Uncovering student ideas in science: Another 25 formative assessment probes,* volume 3. Arlington, VA: NSTA Press.

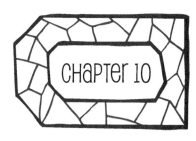

Keeley, P., F. Eberle, and L. Farrin.2005. *Uncovering student ideas in science: 25 formative assessment probes*, volume 1. Arlington, VA: NSTA Press.

Keeley, P., F. Eberle, and J. Tugel. 2007. *Uncovering student ideas in science: 25 More formative assessment probes*, volume 2. Arlington, VA: NSTA Press.

May, K., and M. Kurbin. 2003. To heat or not to heat. *Science Scope* 26 (5): 38.

references

American Association for the Advancement of Science (AAAS). 1993. *Benchmarks for science literacy:* New York: Oxford University Press.

Hazen, R., and J. Trefil. 1991. *Science matters: Achieving scientific literacy.* New York: Anchor Books.

National Research Council (NRC). 1996. *National science education standards.* Washington, DC: National Academies Press.

Watson, B., and R. Konicek. 1990. Teaching for conceptual change: Confronting children's experience. *Phi Delta Kappan* 71 (9): 680–685.

CHAPTER 11

COOL IT, DUDE!

Rosa and Paula stopped in at their favorite sub shop for a sub and a cold drink. As Rosa filled her cup with ice she complained to her friend John, the counter boy. "Why do you guys only give us crushed ice in our drinks instead of ice cubes? The ice melts so fast, we get watered down drinks instead of the real thing."

John rolled his eyes. If he had a quarter for every time that question was asked, he could quit this job and retire.

"The boss says crushed ice cools the drinks off faster," John mumbled.

"What did you say, John boy?" teased Rosa.

"I said," John said slowly and clearly, "the boss says crushed ice cools the drink quicker than cubes."

"Is that really true?" asked Rosa. "Sounds weird. Ice is ice, cubes or crushed. What difference does the size of the ice make?"

Paula piped in, "Well, I do believe crushed ice melts faster, but it waters down the drink faster too!"

"Yeah but people usually drink it fast when they eat subs so it doesn't take them long to empty the cup," replied John. "They want their drink to be cold, fast!"

"I'm not sure I believe that stuff about cooling drinks off faster with crushed ice. You want to prove that to me?" said Paula.

"I don't have to," said John. "I only work here. Find out one way or another for yourself."

"Okay we will," said Rosa, "as soon as we get home."

"And another thing, why use crushed ice? Cause it takes up more space and we get less drink?"

John couldn't let that one go by. "It doesn't make any difference, crushed or cubes, takes up the same space.

"Wait a sec," Rosa responded, " you mean to tell me, if I crushed a cube of ice, it would take up the same space as the cube, uncrushed?"

"Well, I think so," answered John. "At least that's what the boss says. Actually, it doesn't really make a lot of sense, come to think of it. Seems like there *is* more of it, crushed."

The three looked at each other for a moment. Paula finally said, "It has to take up more space, there are more pieces."

John thought a moment. "Well, it's still the same cube—just smaller pieces—we didn't add anything."

"Yeah, but each tiny piece takes up space and there are more ice pieces to fill the cup," countered Rosa.

"Well, it looks like you have two things to prove, now " said John as he wiped off the counter. "Be sure and let me know so I can clue the boss in and get fired!" he said with a grin.

PURPOSE

There are two concepts at work here: conservation of matter and the question about many surfaces vs. fewer surfaces absorbing heat. You may wonder what this story is doing in the Earth system science area, but it has to do not only with thermodynamics and conservation of matter but with water, arguably the most important material we have on the planet. You may think it sounds more like a physics story, but I suspect that more integration of the various sciences is done in Earth science than in any other of the disciplines because it uses so many concepts from other areas in order to understand its overarching view of the world we live in. I also feel that we do too much compartmentalizing of disciplines so that students often do not see how they all fit together. In any case, Rosa and Paula shouldn't have trouble finding out some answers to their dilemmas in their own kitchens, and nor should your students.

RELATED CONCEPTS

- Conservation of substance
- Heat absorption and surface area
- Water
- Displacement
- State changes
- Solids and liquids

DON'T BE SURPRISED

There are several possible things your students may have already formed misconceptions about that you might want to think about as you use this story. Young children may believe that if you break something into pieces, there is more of it than when it was whole. Students usually understand the flaw in this kind of thinking on their own when they are ready, but helping them see that the mass of things does not change when their shape is modified or broken into pieces is not time wasted.

The other misconception may exist into adulthood and may be semantic as much as it is scientific: Many people believe that cold moves out of ice cubes into the drink and thus cools it. This is contrary to the scientific view that *thermal energy* moves from the warmer to the cooler. In fact, *heat* is defined as a transfer of energy from an object that is hot to one that is cooler. You may want to suggest that they think of thermal energy as the mover and of cold as a lack of thermal energy. It is the warmth that does the moving, not the cold. I will address this in more detail in the content background section. It is imperative that they think of energy transfer in this way if they are to make sense out of the story and the cooling of a drink by use of ice.

CONTENT BACKGROUND

Have you ever seen children break cookies into pieces so they have more to eat? Well, they may do this until they have developed beyond what Piaget calls the *preoperational* stage of development. Then they realize that if more was not added or taken away, the amount is the same, whole or broken. This is important in this story if one is comparing the mass of an ice cube as a whole to the mass of the ice cube after it has been crushed. If your students take two identical ice cubes, crush one and leave the other whole, their masses will remain the same. They can verify this on a balance or by melting the two setups and comparing the amounts of water in each. Students will need to be satisfied that both ice cubes are the same size to begin with.

But, if your students wonder if a cupful of crushed ice by volume is the same as a cupful of whole ice cubes, it turns out to be a different story. Think of it this way: because of the shape of the cube, a cupful will have a great deal of air space between cubes. With crushed ice, the particles are smaller and fit more closely together leaving little space between each piece of ice. It seems logical that if the cup sizes are equal, more crushed ice will fit into a cup than ice cubes. One way to show this is to fill a cup with crushed ice and another cup of the same size with cubes. Allow them to melt and you will find more water in the crushed ice cup than the cube cup. So, it seems that if you are dispensing a drink and fill your cup with crushed ice, you will have less room for the drink than if you filled your cup with cubes.

The question then becomes, how much crushed ice do you put into your cup so that you are not cheating yourself on the amount of drink you purchase? This can only be decided by how fast you wish to cool your drink and how fast you drink it. There is no formula, because all ice crushers are different and therefore the particle size is different. The best answer is probably that since ice is free, you can always go back to the dispenser and replenish your ice but not your drink. So if you put a small amount of crushed ice into the cup, and your drink is not cold enough after a few minutes, you can always add more ice.

So, about the other question in the story—will crushed ice cool the drink faster? Heat transfer takes place at the interface of the surfaces of the cooler object. Let's say that we have an ice cube that is 3 cm². Since there are six sides to the cube, there is approximately 54 (3 cm × 3 cm × 6 sides) square centimeters of surface area touching the liquid. If you were to cut up the same full-size cube into 1 cm cubes, there would be 27 little cubes, each with a surface of 1 cm². Note that there is still the same amount of ice but it is now broken up into smaller pieces with more surface area to interact with the liquid. So 27 (cubes) × 6 (sides) × 1 gives you 162 cm² of surface area, or three times that of a whole cube. It is no wonder that with the additional surface area interacting with the liquid, heat would be absorbed more quickly. But, with the quicker cooling the ice is melting more quickly, diluting the drink as Rosa complained. It is a dilemma! For the math and the idea for showing the increase in surface area I thank the Worsley School in Alberta, Canada, and their website, which is full of great ideas

NATIONAL SCIENCE TEACHERS ASSOCIATION

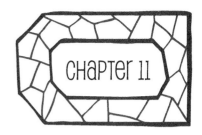

for teachers on all subjects: *www.worsleyschool.net/science/sciencepg.html*. Note that it is true the ice crusher does not crush the ice into nice little cubes, but it does create more particles with more surface area to interact with the drink and absorb the heat.

Heat and temperature are two entirely different things. Heat is commonly referred to as the amount of *energy* in a substance, and the measure of this for everyday purposes is by a thermometer, which determines the average amount of heat in a substance or body. Temperature is a human-devised concept set up on arbitrary scales such as Fahrenheit, Celsius, or Kelvin.

Every substance has some heat in it unless it has somehow miraculously reached the temperature of absolute zero, a temperature almost impossible to attain even in a laboratory. Absolute zero is reached when no more heat energy can be extracted from a substance. The larger the substance, the more heat is present. Two ice cubes have twice the heat energy as one ice cube. A seemingly puzzling fact is that the water in an almost-frozen swimming pool at 3°C has more heat in it than a glass of water at a temperature of 100°C merely because there is a great deal more water in the pool, and the amount of potential and kinetic energy in any substance is directly related to the amount of the substance.

Heat energy is attributed to the motion of atoms in any substance. More atomic activity means greater heat and less activity means less heat. So, when you heat or cool something, you are changing the activity level of its atoms. Also, heat energy can be transferred from one substance to another. By adding energy, the amount of heat a thing contains increases. A heat donor, such as the Sun, electricity, burner, or nearby higher energy source transfers its energy to the heat receiver. This is essentially the first law of thermodynamics.

Heat transfer, from the warmer to the cooler, can be done by one of three methods: *conduction, radiation,* or *convection.* You have felt the result of *conduction* when you put a spoon into a hot cup of liquid and then touched the spoon. The heat energy is transferred directly from the collision of the atoms in the liquid to the atoms in the spoon to you. You may also have felt the transfer of energy by *radiation* if you stood close to a fire, an electric heater, or a lamp. The energy of the heat source is in the form of infrared energy (a part of the light spectrum), which in turn excites your heat sensors and you feel heat. In *convection*, the atoms in a liquid or gas set up a current of rising and falling atoms that eventually bring everything in the substance to the same temperature. An interesting phenomenon about conduction is that some substances conduct heat better than others. For instance, if you touch metal, it feels cooler than other substances in a room. This is because the heat from your body transfers more quickly to the metal and it feels cooler to you. If the metal has been in the room for a long time, it will have the same temperature as the rest of the objects in the room. Your body will be fooled into thinking that the metal is cooler when it is really not.

Many students live with a common misconception that cold is a form of energy that can move from one place to another. They may also reckon that there is an unlimited supply of this "cold" in the ice that can continue to move into the drink and drop the temperature until the ice is gone. In their minds, the "cold" in the ice disappears into the drink until it is all used up. If one believes this, it is entirely possible for the drink to become colder than the temperature of the ice itself. We know this to be untrue since the heat of the drink will cause the ice to melt rather than allowing the ice to continue to decrease the temperature of the drink.

This phenomenon can be tested with the aid of a thermometer and a glass of ice water. The heat in the drink changes the phase of the ice from solid to liquid by increasing the energy in the atoms in the ice. This will eventually result in temperature equilibrium between the ice and the drink. When equilibrium has been reached, the temperature will go no lower because there can be no further flow of energy since everything is the same.

With crushed ice, the heat transfer will be faster because of the amount of surface area, but, again, once equilibrium is reached the drink will become no cooler. Because the ice is crushed and has more surface area, this will occur more quickly than it would with whole ice cubes.

Your students should have a great time arguing this one and in the process they will be dealing with the properties of water in its various states and with the laws of thermodynamics as applied to the cold drink dispensers so prevalent in their daily lives.

related Ideas From The National Science Education Standards (NRC 1996)

K–4: Properties of Objects and Materials
- Materials can exist in different states—solid, liquid, and gas. Heating or cooling can change some common materials, such as water, from one state to another.

K–4: Light, Heat, Electricity, and Magnetism
- Heat can be produced in many ways such as burning, rubbing, or mixing one substance with another. Heat can move from one object to another.

5–8: Transfer of Energy
- Energy is a property of many substances and is associated with heat, light, electricity, mechanical motion, sound, nuclear energy, and the nature of a chemical change. Energy is transferred in many ways.
- Heat moves in predictable ways, flowing from warmer objects to cooler ones until both reach the same temperature.

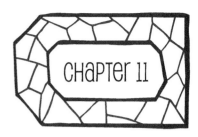
related ideas from Benchmarks for science Literacy (aaas 1993)

K–2: *The Structure of Matter*

- Heating and cooling cause changes in the properties of materials. Many kinds of changes occur faster under hotter conditions.

3–5: *Energy Transformations*

- When warmer things are put with cooler ones, the warm ones lose heat and the cool ones gain it until they are all the same temperature. A warmer object can warm a cooler one by contact or at a distance.

6–8: *Energy Transformations*

- Heat can be transferred through materials by the collision of atoms or across space by radiation. If the material is fluid, currents will be set up in it that aid the transfer of heat.
- Energy appears in different forms. Heat energy is in the disorderly motion of molecules.

USING THE STORY WITH Grades K–4

If it makes more sense to your younger students to change the characters in the story to children and their parents, you may do so easily without changing the nature of the dilemma. True to the basic premise behind this book, this story focuses on some of the things in our lives that often go unnoticed and yet have a significant amount of science behind them.

You can get an idea of what your students are thinking about the topic by using one or both of the probes in *Uncovering Student Ideas in Science*, volumes 1 and 2 (Keeley, Eberle, and Farrin 2005; Keeley, Eberle, and Tugel 2007). "Ice Cubes in a Bag" is in volume 1 and "Ice Cold Lemonade" is in volume 2. In "Ice Cubes in a Bag," students are asked to decide if ice gains or loses mass when it melts. In "Ice Cold Lemonade," they are asked what they think about "coldness" and "heat." Both probes will evoke discussion and allow you to decide if they are ready to discuss the story. "Ice Cubes in a Bag" is best done with the ice in a bag as a prop and putting it on a balance to see if anything changes as the ice melts. Of course it doesn't. This allows you to try the investigation in several ways if the students are not convinced. They might want to use more ice cubes or just repeat the investigation.

The "Ice Cold Lemonade" probe might be difficult for younger children to fathom since the terms in the probe ask students to distinguish between heat and cold, while hot and cold are really just parts of a continuum. We often tell people to close the door to "keep the cold out," when we really should be saying, "keep the heat in." It's an example of sloppy language that helps to fuel misconceptions. For young students it is just a matter of words, but to a scientist, it is the first law of thermodynamics and inviolable. Heat energy moves from the warmer to the cooler. Understanding this is paramount to grasping many of the concepts revolving around energy transfer. But for young children, trying out the investigation in the story without attempting the development of the scientific concepts might be well enough in itself and can provide supporting beams in building the scaffolding toward later understanding of energy transfer.

Third and fourth graders are very capable of reading thermometers and of setting up "fair" investigations regarding the arguments set forth in the story. It cannot be overstated that students of this age need to be reminded about conducting "fair" investigations. Variables and their control should be considered every time a new experiment is conducted. Students of this age need constant reminders about making sure that all variables except the one being tested need to be kept the same. Just recently I had the opportunity to interview a 10-year-old student about what a "fair" experiment meant to her. Even though she had just been reminded about variables and fairness in an interactive investigation, she replied that a "fair experiment" was one where she got the results she predicted. It is an enigma to me that some children are the first to point out unfairness in games, yet find it difficult to transfer this idea to investigations—even ones that seem similar to games. It is our role to keep reminding students about this part of research although it may seem to us repetitive and unnecessary.

Since most third and fourth graders have had the experience of using the soft drink dispensers, they are ready to test the questions asked by the story. Many will be surprised by how much difference there is in the cooling time between crushed ice and cubes and most are very surprised that so much more crushed ice fits into a drink cup. You may add an embedded assessment here and ask them how they are going to prepare their next drink at the sub shop. They can write this anecdotally in their science notebooks and this will give you a good idea about how they are applying their new knowledge.

I have had great success with an ice cube–melting race at this grade level. Students are given an ice cube and told to do anything they can to melt it except touch it with any part of their bodies. They often explain afterward that they had to get "heat" into the cube to melt it. When this happens, they are on their way to accepting that ice changes state when heat is added. I would not even attempt to distinguish between temperature and heat. Leave that for the next level.

USING THE STORY WITH GRADES 5–8

The story itself should elicit a myriad of opinions since middle schoolers have had lots of experience with soft drink machines. This can lead to consumer-based investigations on their part. If you are able to have a blender on hand to offer a

ready supply of crushed ice, you and your students will have a lot of fun developing various scenarios to find the best way to get your money's worth at the dispenser. Plenty of thermometers and cups should be available. We have found that the "micro" approach is the most economical, since you can get the same result with small cups as you can with large ones and use less ice in the process. If possible, use insulated cups so that the students' handling of their setups won't affect their results. It is a good idea to separate the two quandaries at first and have the students come up with two separate problem statements and go from there to design investigations to test their predictions. Don't forget to insist on reasons for their predictions based on some previous experience or scientific basis. This should be in their science notebooks for you to check per your scheduled timetable.

Since the National Science Education Standards suggest that trying to get your students to truly understand the difference between temperature and heat is a waste of time and effort, I suggest that you give this area minimal attention unless you are so required by your local standards. Time might be better spent having students learn to graph their data and use these data to draw conclusions and apply them to real-life situations. Does it make more sense to fill your cup with drink and then add ice until the desired temperature is reached? What kinds of problems arise with this method? Should one add ice first to a low level and then test the drink and go back for ice? What kinds of problems arise with this method? If you have access to the digital probe thermometers and the computer programs that help with the graphing and accumulating of data, your students will have even more data to display in their presentations.

I have found that if various groups come up with different tests and bring their investigation designs to the whole class to critique, everybody gets to be involved in all of the investigations to some extent. Then the groups can take the suggestions and carry out the investigations, knowing that the entire scientific community has agreed on the methodology. This helps with the sharing of data and conclusions. It can also be another embedded opportunity for formative assessment for you. Of course, this takes more time but the group construction of knowledge is well worth it.

Disagreement is inevitable and in fact should be encouraged since it is during discussions that the main conceptual understandings are built. Students should be encouraged to engage in constructive argument since this is the way that the scientific community comes to agreement. Always take this opportunity to help your students see that science is based on evidence and not opinion.

Read the article "Teaching for Conceptual Change" by Watson and Konicek (1990). This chronicles the experience of an elementary teacher whose students believe that a mitten or any wool garment produces heat. It intersperses the philosophy of inquiry teaching with the actual documentation of the teacher's attempts to allow her students to test their misconceptions against the reality of the day-to-day investigations as dictated by their questions.

related BOOKS anD NSTa JOURNaL arTICLes

Ashbrook, P. 2006. The matter of melting. *Science and Children* 43 (4): 19–21.

Damonte, K. 2005. Heating up and cooling down. *Science and Children* 42 (8): 47–48.

Driver, R., A. Squires, P. Rushworth, and V. Wood-Robinson, 1994. *Making sense of secondary science: Research into children's ideas.* London and New York: Routledge Falmer.

Keeley, P., F. Eberle, and L. Farrin. 2005. *Uncovering student ideas in science: 25 formative assessment probes,* volume 1. Arlington, VA: NSTA Press.

Keeley, P., F. Eberle, and J. Tugel. 2007. *Uncovering student ideas in science: 25 more formative assessment probes,* volume 2. Arlington, VA: NSTA Press.

Line, L., and E. Christmann, 2004. A different phase change. *Science Scope* 28 (3): 52–53.

May, K., and M. Kurbin. 2003. To heat or not to heat. *Science Scope* 26 (5): 38.

Pusvis, D. 2006. Fun with phase changes. *Science and Children* 29 (5): 23–25.

Robertson, W. 2002. *Energy: Stop faking it! Finally understanding science so you can teach it.* Arlington, VA: NSTA Press.

references

American Association for the Advancement of Science (AAAS). 1993. *Benchmarks for science literacy.* New York: Oxford University Press.

Hazen, R., and J. Trefil. 1991. *Science matters.* New York: Anchor Books

Keeley, P., F. Eberle, and L. Farrin. 2005. *Uncovering student ideas in science: 25 formative assessment probes,* volume 1. Arlington, VA: NSTA Press.

Keeley, P., F. Eberle, and J. Tugel. 2007. *Uncovering student ideas in science: 25 more formative assessment probes,* volume 2. Arlington, VA: NSTA Press.

National Research Council (NRC). 1996. National science education standards. Washington, DC: National Academies Press.

Watson, B., and R. Konicek. 1990. Teaching for conceptual change: Confronting children's experience. *Phi Delta Kappan* 71 (9): 680–685.

Worsley School. Science and mathematics. *www.worsleyschool.net/science/sciencepg.html.*

CHAPTER 12

THE NEW GREENHOUSE

Eddie and Kerry's mom is a master gardener. She takes care of other peoples' gardens and raises plants from seeds and from what she calls cuttings to sell to her customers. This year she had a new building installed in the yard to help her expand her business. It was called a solar greenhouse. It was about 8 feet (2.5 meters) wide, 10 feet (3 meters) long, and about 10 feet (3 meters) high at the peak. It looked like a one-story house with glass walls and glass roof, two windows up on the top that could be raised or lowered, and a door. It took a lot of time and a lot of people to put it together, but when it was finally up, it looked great. They put in a water hose and electrical outlets. Eddie and Kerry's father built a bunch of tables inside with screen tops so that Mom could water her seedlings, letting the water seep through onto the gravel on the floor.

In the early spring of the year, their mother planted a lot of seeds in little trays and put them in the greenhouse. She had never had a greenhouse before and had a lot to learn about how they work. She knew that the Sun was supposed to shine down through the windows and warm the inside even though it was chilly outside. Somehow, the heat that came in through the windows stayed inside the greenhouse and didn't go back out. The windows were sealed, the door was tight, and the Sun shone brightly on the greenhouse. During the day in an unusual spring warm spell, the temperature got so hot that the windows and doors had to be opened to keep the temperature from frying the little seedlings. She even put a fan in the door to bring in some cool air to keep the temperature at a reasonable level.

Evening and overnight was a different matter however. As the temperature outside dropped and the Sun set, the temperature in the greenhouse was no more than a few degrees warmer than the outside air and, in early spring in New England, that was near or at freezing. Those little seedlings were in trouble.

"I am at a loss about how to keep the climate in my new greenhouse at a level that is good for the plants!" said Mom.

"Why don't you look on the internet and find out about greenhouse care?" said Kerry.

That's exactly what she did. She found out that in her case, she was dealing with what was called a passive solar greenhouse, or a greenhouse that was strictly at the mercy of the Sun. Commercial greenhouses had heaters and sprinklers, automatic vent raisers, and all sorts of gadgets much too expensive for Mom's business. One article suggested putting large rocks in the greenhouse to soak up heat during the day and give it back off during the night.

"Where am I going to get huge rocks and if I get them, where am I going to put them so there is still room for the plants?"

"Maybe we can use something other than rocks to collect heat," said Eddie. "I read that different things take in heat and give it off, either quickly or slowly. Don't remember which ones though. I also heard that dark colors make a difference, but I am not so sure that makes sense."

"I guess we can try some things. I have buckets for water or some other liquid we could try. Maybe that would work," said Mom.

"But water is cold. Do we have to use warm water?" Eddie asked.

"Well, I don't really know. But I do know it's going to take a lot of experimenting and keeping records," said Mom. "Everybody up for that?"

PURPOSE

In our economy, with the recent oil crisis and the problem of fuel supply versus insatiable demands, the importance of alternative energy sources is abundantly clear to anyone who has been to a gas pump or seen an electric bill lately. Those of us concerned about the health of the planet are sensitive to the differences between energy sources and how we use them and the disparity in this among poor and wealthy countries. Young children may be aware only of what they hear from adults, while older children may even be involved in organizations that are trying to make these problems known to the public. All of us, young and old, are now conscious of the differences it is making in our lifestyles, be it in concerns about automobiles and miles-per-gallon ratios or the need to take fewer trips due to gas costs. Developing an alternative energy project can make all of us more aware of the finite aspects of the current sources of power.

The technological aspects of this story are also important in that students can use their new learning about energy supply and storage for the development of a more efficient product. This is putting science to use in solving a real-life problem.

RELATED CONCEPTS

- Technology
- Renewable energy
- Energy transformation
- Conservation of energy
- Alternative energy
- Solar energy
- Thermodynamics

DON'T BE SURPRISED

Your students may not realize that the Sun's rays differ in their angle during the year. Many will still believe that the reason for summer and winter is the distance of the Earth from the Sun. They do not realize that it is the angle of the Sun striking the Earth that causes seasons. Yet the angle and direction of the Sun during the day is important in heating solar greenhouses. Many students are also not aware of the differing heat absorption and radiation properties of materials. They may be unaware of the terrific amount of heat that can be captured by a transparent building and be amazed by the great disparity between outdoor and indoor temperatures on sunny days. Some may even be surprised that solar energy can be "stored" in substances like water or rocks.

CONTENT BACKGROUND

This story is based upon using solar energy and a variety of materials to modify and channel this energy to capture and hold heat. Almost everybody has

experienced the differences in temperature due to sunshine passing through windows into an enclosed space. Solar light is composed of many wavelengths, ranging from the subspectrum invisible level through the visible spectrum and beyond to invisible ultraviolet ranges and above. This means that light comprises low-energy to high-energy ranges. Scientists call this *electromagnetic (EM) radiation.* In some cases this radiation behaves like waves and in others, like particles. (Quantum mechanics has attempted to unify these two models of EM but we won't go there in this book! Let's just say that scientists have agreed that science can live with seemingly contradictory explanations and that sometimes it makes sense to explain EM radiation as wave behavior and sometimes as particle behavior.) In talking about solar energy, it is most useful to talk about EM radiation as waves. There are radio waves that have low energy, and gamma rays that have high energy. In between, from low to high, are microwaves, infrared waves, our visible spectrum, ultraviolet waves, and x-rays.

When light from the Sun enters an enclosed area by way of a transparent passageway, such as a window or transparent or translucent plastic, it enters in a short-wave and high-energy state. Objects in the space, such as dirt, wood, and water, absorb it. The light is then reradiated from these objects as lower-energy waves which do not readily pass out of the clear areas through which the light entered. In essence, it is "trapped" in the enclosed space, and that space heats up as a result. This is the famous "greenhouse effect." We are all familiar with it if we have ever stepped into a car that has been sitting in the sun with the windows rolled up. It can be cool outside, but the car is much warmer than the outside air.

According to some experts, if you have in the greenhouse approximately two gallons of water for every square foot of glazing, you can maintain a temperature about 30°F above the outside temperature. Massive material, like rocks or concrete helps greatly, but pound for pound, water in closed, black containers is the best retainer of heat and the best way to keep temperatures in greenhouses stable. The reason for this is that water is slow to store heat and just as slow to transfer it. Different substances have what is known as *specific heat*, meaning they absorb and transfer heat at different rates. Metals need less heat to raise their temperatures and transfer heat very quickly as well. Rock or concrete needs more heat to raise its temperature, and water needs the most. The concrete paving and buildings in cities help keep city streets warm on sunny days, and the oceans and lakes have a moderating effect on temperatures in seaside or lakeside communities.

Our experience in our passive solar greenhouse is that ventilation is necessary on very sunny days since the temperature can reach over 100°F by late afternoon, which will likely kill the plants inside. So vents in the roof allow the warm air to escape into the cooler atmosphere. A fan is often also necessary to keep the temperature from rising to dangerous levels.

All of this is pertinent to the greenhouse effect so prevalent in the news today, since many scientists and climatologists believe that our society is producing what are known as "greenhouse gases" in such great amounts that we are in essence preventing heat energy from escaping from the Earth as it used to, thus causing

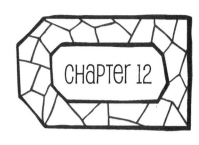

the temperature of the atmosphere to rise. The gases are forming a barrier in the atmosphere, just like the glazing in a greenhouse or the glass in a car, and preventing the release of heat back into space. Thus the average temperature of the Earth is rising. The term for this is, of course, *global warming*.

With this story, students can explore the effects of solar energy, energy absorption, and energy dissipation of various substances and also the effect of dark and light colors on the absorption of heat in closed systems.

related ideas from the National science education standards (Nrc 1996)

K–4: *Objects in the Sky*
- The Sun provides the light and heat necessary to maintain the temperature of the Earth.

K–4: *Abilities of Technological Design*
- In problem identification, children should develop the ability to explain a problem in their own words and identify a specific task and solution related to the problem.

K–4: *Abilities of Technological Design*
- Students should make proposals to build something or get something to work better; they should be able to describe and communicate their ideas. Students should recognize that designing a solution might have constraints, such as cost, materials, time, space, or safety.

5–8: *Transfer of Energy*
- Energy is a property of many substances and is associated with heat, light, electricity, mechanical motion, sound, nuclei, and the nature of a chemical. Energy is transferred in many ways.
- Heat moves in predictable ways, flowing from warmer objects to cooler ones, until both reach the same temperature.
- The Sun is a major source of energy for changes on the Earth's surface. The Sun loses energy by emitting light. A tiny fraction of that light reaches the Earth, transferring energy from the Sun to the Earth. The Sun's energy arrives as light with a range of wavelengths, consisting of visible light, infrared, and ultraviolet radiation.

5–8: *Understanding About Science and Technology*
- Science and technology are reciprocal. Science helps drive technology as it addresses questions that demand more sophisticated instruments and

provides principles for better instrumentation and technique. Technology also provides tools for investigations, inquiry, and analysis.

related ideas in benchmarks for science literacy (aaas 1993)

K–2: Design and Systems
- People can uses objects and ways of doing things to solve problems.

K–2: Energy Transformations
- The Sun warms the land, air, and water.

3–5: Design and Systems
- There is no perfect design. Designs that are best in one respect may be inferior in other way. Usually some features must be sacrificed to get others.

3–5: Energy Transformations
- Things that give off light often also give off heat.
- Some materials conduct heat much better than others. Poor conductors can reduce heat loss.

6–8: Design and Systems
- Design usually requires taking constraints into account. Some constraints, such as gravity or the properties of the materials to be used, are unavoidable.

6–8: Processes That Shape the Earth
- Human activities, such as reducing the amount of forest cover, increasing the amount and variety of chemicals released into the atmosphere, and intensive farming, have changed the Earth's land, oceans, and atmosphere. Some of these changes have decreased the capacity of the environment to support some life forms.

6–8: Energy Transformations
- Energy cannot be created or destroyed, but only changed from one form into another.
- Most of what goes on in the universe—from exploding stars and biological growth to the operation of machines and motion of people—involves some form of energy being transformed into another. Energy in the form of heat is almost always one of the products of an energy transformation.
- Energy appears in different forms. Heat energy is in the disorderly motion of molecules.

USING THE STORY WITH GRADES K–4

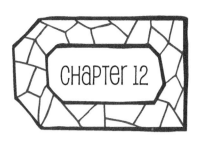

With thanks to Peggy Ashbrook (2007), I suggest that very young students begin their inquiry into this story outdoors on a sunny day with a discussion about what they can sense about the light coming from the Sun. Moving the students around from an absorbing surface like asphalt to more reflective surfaces may let them feel the differences in the heat. Making "Sun prints" on light-reactive paper using incandescent light and sunlight and then comparing them may give the children a sense of the power of the Sun's energy and its effect upon them and the world around them. Connecting this to the story can be done by asking students to relate how the energy from the Sun could cause the greenhouse to heat up. Since the students have developed an appreciation for the power of the Sun's energy, they may well be able to relate the two situations but not much more at this age.

When working with grades 3 and 4, I believe it would be advantageous to focus on the absorbing function of dark colors and the reflecting function of light colors before considering whether or not to address the greenhouse effect, which is the basis of the story. This can be done by placing thermometers on dark surfaces and on light surfaces and comparing temperatures. Ask them if they have noticed that the color of clothes they wear on really sunny days affects how hot they feel. This should lead to a discussion about the amount of solar energy the dark-colored clothing absorbs or light-colored clothing reflects. Then, investigations can begin with thermometers and various colors of T-shirts or baseball caps.

Once you have determined if the students are aware of the absorbing and reflective attributes of objects, they may be able to move into the greenhouse and realize that the soil and other objects in the greenhouse have been able to absorb the Sun's energy and hold it inside the structure. A field trip to a local botanical garden may be in order. It would be surprising if the students at this level were able to apply their knowledge of solar energy to the technological aspects of the greenhouse. But it is possible, especially with older students or ones who may have had experience with enclosed solar rooms, either in their own homes or in family businesses. Read the following segment on using the story with grades 5–8 to see if your students are up to the complexities of investigating some of the questions that arise.

I also recommend you read the Damonte article in *Science and Children,* "Heating Up and Cooling Down" (2005), which will be very appropriate, even for younger children.

USING THE STORY WITH GRADES 5–8

This story lends itself to a great deal of scientific and technological investigation for children at this grade level. Gregory Childs, in his *Science and Children* article "A Solar Energy Cycle" (2007), suggests checking out whether or not your

students are aware of the absorbing and reflecting power of light and dark surfaces first before moving into the greenhouse investigations. He uses two paper houses, one black and one white, placed in the sun to compare temperatures. They need not be works of art—simple cardboard boxes should do nicely. The results of this investigation will be an imbedded formative assessment for you to see what your students understand about this concept.If you are comfortable with what you find, you can move on to the ideas involved in the story about the greenhouse effect.

The story itself gives clues to the differences in heat absorption of various materials, and controlled experiments can be conducted using greenhouses constructed by your students using cardboard boxes and plastic wrap or more authentic balsa wood greenhouse constructions, depending upon the sophistication of your students. Many students have built greenhouses in the 12" × 14" size, many complete with vents in the upper windows using duct tape for hinges, an so on. My own attempt at a very large plastic solar greenhouse built by my class on the roof of my school was thwarted when, during a wind storm, it became a wonderfully aerodynamic box kite, sans line, and ended up several blocks away from the school in someone's backyard! Luckily no property or bodily damage resulted, and the recipients of the structure were friendly and understanding. But I learned a great deal about size, wind, and the importance of anchoring. For this reason, I do suggest keeping your projects small and portable so that they can be taken indoors and out of the unpredictable elements.

There are a few variables to be tested here, for example, the effect of dark and light surfaces in the greenhouse; the differences in storage mass material; the effects of insulation; and most basic of all, witnessing the simple solar greenhouse effect in order to establish a baseline set of data. You may be able to round up some of the black plastic film canisters that, with lids, make wonderful water, rock, sand, or gravel containers. It is good to remember that we are talking about mass here, and controlling the mass is important when comparing the various materials.

On sunny days, exposures for 20 minutes are usually sufficient, with a reading every 2 minutes or so. This gives a great deal of data to be graphed and analyzed. If you are fortunate to have a temperature probe and graphing program, this can be used as well. If your students can procure digital probe thermometers, it makes reading easier, but if not, a thermometer placed near a window of the greenhouse can be read without opening the structure. You can also test color absorption and temperature effects by using different color floors in the greenhouses. This has proved to be very effective in producing significant data. With guidance, your students will be able to make a list of significant variables that must be controlled and the results should be rewarding both from a conceptual understanding and a technological standpoint.

Of course, no project on solar energy is complete without a discussion of alternative energy. With oil and gas prices fluctuating wildly and these fossil fuels being depleted, solar energy is poised to become a major alternative. Some schools have hosted simulated symposia and/or town hall meetings to discuss this issue. The classroom setting can be very conducive for impressionable students to become aware of the seriousness of this issue as it relates to their future.

related BOOKS aND NSTa JOURNAL articles

Ashbrook, P. 2007. The early years: The sun's energy. *Science and Children* 44 (7): 18–19.

Childs, G. 2007. A solar energy cycle. *Science and Children* 44 (7): 26–29.

Damonte, K. 2005. Heating up and cooling down. *Science and Children* 42 (8): 47–48.

Driver, R., A. Squires, P. Rushworth, and V. Wood-Robinson. 1994. *Making sense of secondary science: Research into children's ideas.* London and New York: Routledge-Falmer.

Keeley, P. 2005. *Science curriculum topic study: Bridging the gap between standards and practice.* Thousand Oaks, CA: Corwin Press.

Keeley, P., F. Eberle, and C. Dorsey. 2008. *Uncovering student ideas in science: Another 25 formative assessment probes*, volume 3. Arlington, VA: NSTA Press.

Keeley, P., F. Eberle, and L. Farrin. 2005. *Uncovering student ideas in science: 25 formative assessment probes*, volume 1. Arlington, VA: NSTA Press.

Keeley, P., F. Eberle, and J. Tugel. 2007. *Uncovering student ideas in science: 25 more formative assessment probes*, volume 2. Arlington, VA: NSTA Press.

May, K., and M. Kurbin. 2003. To heat or not to heat. *Science Scope* 26 (5): 38.

references

American Association for the Advancement of Science (AAAS). 1993. *Benchmarks for science literacy.* New York: Oxford University Press.

Ashbrook, P. 2007. The early years: The sun's energy. *Science and Children* 44 (7): 18–19.

Childs, G. 2007. A solar energy cycle. *Science and Children* 44 (7): 26–29.

Damonte, K. 2005. Heating up and cooling down. *Science and Children* 42 (8): 47–48.

Hazen, R., and J. Trefil. 1991. *Science matters: Achieving scientific literacy.* New York: Anchor Books.

National Research Council (NRC). 1996. *National science education standards.* Washington, DC: National Academies Press.

WHERE DID THE PUDDLES GO?

It rained very hard all night. Aunt Sophia said it rained "cats and dogs," but Vashti didn't see any on the ground, so she thought Aunt Sophia must be kidding. Aunt Sophia was a great kidder. You never knew when she was serious.

But Vashti did see lots of other things on the ground, like leaves, branches, and, most of all, puddles.

Puddles were everywhere. Wherever there were holes or depressions in the ground, there were puddles.

The puddles were on the street and the sidewalk; many looked dirty and muddy.

Now don't think that Vashti was a stranger to puddles. They were nothing new to her. In fact, when she was very little, she loved to put on her boots and splash in them, often getting herself awfully dirty as the water and mud splashed up onto her clothes. But, that was when she was little and she didn't do that anymore; although sometimes, like this morning on her way to school, they looked mighty inviting. There were still dark clouds overhead and it was windy, and even though she hurried to get indoors, she couldn't help wondering about the puddles. Because she knew some would disappear soon and some would disappear much later.

"Sometimes the big puddles dry up faster than the little ones," she thought to herself as she walked along. "Why would that happen?"

She walked over to a few puddles, thought for a while, then smiled. "You know, I think I know why!"

Just then, her friend Juana came out of her building to join her, and Vashti had an idea for a joke she could play on her.

As they passed the basketball court where the big kids played ball at night, she saw just what she expected, puddles of all sorts all over the court. She walked over to the biggest puddle and said to Juana, "I'll bet you this big puddle will be gone when we come home this afternoon, and this littler one over here won't. Wanna bet?"

"Sure," said Juana. "You think that big puddle will dry out faster than this little one, huh? Okay, you're on. Looks like a no-brainer to me. The rain is over and there're a few patches of blue up in the sky so let's go to school and we'll see who's right this afternoon!"

The girls got involved in some afterschool activities, and it was late afternoon before they began their walk home. As they passed the basketball court, sure enough, Vashti's puddle was all dried up and Juana's little puddle still had some water in it.

"How'd you do that?" asked Juana. "The sun wasn't even out!"

"Piece of cake!" said Vashti.

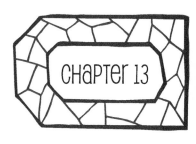
PURPOSE

This story has to do with evaporation. We will explore the major factors that lead to how quickly water evaporates. We have all had the opportunity to observe that there are differences in how quickly water evaporates from puddles after a rain, but perhaps we didn't notice it enough to wonder why. This story focuses, as usual, on what we might have missed.

RELATED CONCEPTS

- Surface area
- Water cycle
- Energy
- Evaporation

DON'T BE SURPRISED

Some students may not believe that it is possible for a puddle that appears larger to evaporate more quickly than one that seems smaller. The trick in Vashti's prediction is that she picks a big puddle that is shallow and not the smaller one that is deeper. The smaller puddle presented less surface area to the atmosphere. Your students may not be aware of the importance of surface area in evaporation or in energy gain and loss.

CONTENT BACKGROUND

The phenomenon of water evaporating is a common one in everyday life. Most texts focus on the water cycle and how the change of state undergoes a cycle from liquid to gas to liquid again. This story is more focused on just one part of the cycle—evaporation—and under what conditions it occurs more readily. It is one of the everyday science mysteries we often ignore, although it is quite interesting when we take note of it.

Since water changes from its liquid state to a gaseous state so frequently in our lives, we often fail to notice under what conditions it dries up the quickest. Dew disappears without our noticing it; water on a bicycle seat is there one minute and gone the next, or so it seems. Vashti is aware of something that Juana is not: Puddles come in all sizes and depths. Water tends to accumulate in any depression since water flows downhill due to the pull of gravity. If the depression into which the water flows is deeper, more water collects. Two things are at play here, the amount of water and the amount of surface of area that is exposed to the air.

Just as surface area plays a part in melting ice or absorbing heat, it plays a large part in the evaporation of water into the air. Lakes and oceans are credited for about 90% of the water found in the atmosphere. These two bodies of water also present the largest surface area to our global atmosphere. Plants provide a

small part of the water through trans-evaporation from their leaves; also, a small amount of atmospheric water comes from a process called *sublimation* in which water molecules actually find enough energy in ice and snow to evaporate without going through the melting process first.

Surface area is important in all sorts of energy exchanges. In chapter 11 of this book, the characters in the story "Cool It, Dude!" discovered that ice cubes that were crushed absorbed heat faster than whole ice cubes. The clue was that there was more surface area available in crushed ice than in the standard ice cube.

Warm water cools faster in a saucer than in a cup. Ice freezes faster in a shallow container than in a cup. These phenomena are due to the differences in surface area. Likewise, the same amount of water in a shallow dish and in a cup will evaporate at different rates. When water evaporates, it is because the molecules of water have enough energy to escape the bonds that they hold to other water molecules. The pressure of the air above has an effect on how easily they escape too. The greater the pressure exerted by the air above, the more difficult it is for the molecules to escape. So, it makes sense that the greater the surface of the water exposed to air, the more space and possibility for molecules to escape from that surface into the air.

Thus, Vashti noticed that some of the puddles were shallow and spread out over the surface of the blacktop. She also noticed that some of the seemingly smaller puddles were made up of accumulated water in deeper depressions. Although Vashti may not have known of the concept of evaporation, she was observant and relied upon her experience to make her prediction.

If you have ever witnessed the making of maple syrup in one of the northern states, you may have noticed that the sap from the maple trees is placed in a shallow pan on top of the fire. To make the syrup, the water—the main ingredient in sap—has to be removed through evaporation, leaving the sugary part behind. It normally takes about 40 gallons of sap to make one gallon of syrup. That's a lot of water to evaporate! The evaporation pan is shallow and large with lots of surface area. Clouds of water vapor fill the sugaring shack with a sweet humidity. Even with a pan designed for quick evaporation, the job takes a long time, since close to 98% of the water has to be released into its gaseous form.

When water evaporates naturally from a body of water, the energy that has been spent to release the molecule into the atmosphere leaves the water a little cooler. People who live in hot, dry climates use this phenomenon as a way to keep their homes cool, in a device known locally as a "swamp cooler." This is a cloth-covered drum rotating in a pan of water that has a fan blowing over it into the house. The water evaporating from the cloth on the drum cools the air, allowing the fan to blow cool but humid air into the home. In climates where the relative humidity is about 10%, the extra humidity does not add to discomfort. Swamp coolers use less energy than air conditioning. An interesting fact is that evaporation doubles for every 10°C in temperature.

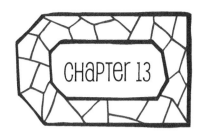

RELATED IDEAS FROM THE NATIONAL SCIENCE EDUCATION STANDARDS (NRC 1996)

K–4: *Properties of Objects and Materials*
- Materials can exist in different states: solid, liquid, and gas. Some materials, such as water, can be changed from one state to another by heating or cooling.

5–8: *Structure of the Earth System*
- Water, which covers the majority of the Earth's surface, circulates through the crust, oceans, and atmosphere in what is known as the "water cycle." Water evaporates from the Earth's surface, rises and cools as it moves to higher elevations, condenses as rain or snow, and falls to the surface where it collects in lakes, oceans, soil, and in rocks underground.

RELATED IDEAS FROM BENCHMARKS FOR SCIENCE LITERACY (AAAS 1993)

K–2: *The Earth*
- Water left in an open container disappears but water in a closed container does not disappear.

3–5: *The Earth*
- When liquid water disappears, it turns into a gas (vapor) in the air and can reappear as a liquid when cooled, or as a solid if cooled below the freezing point of water. Clouds and fog are made of tiny droplets of water.

3–5: *The Structure of Matter*
- Heating and cooling cause changes in the properties of materials. Many kinds of changes occur faster under hotter conditions.

6–8: *The Earth*
- Water evaporates from the surface of the Earth, rises and cools, condenses into rain or snow, and falls again to the surface.

USING THE STORY WITH GRADES K–4

Even though the Benchmarks say that water disappears, we want children to know that water merely disappears *from view* as it becomes vapor. Even very young children have seen liquid in a jar or in puddles disappear from view, but having them believe that the water has changed to vapor and still exists in the air is another story. However, we can still allow our students to participate in investigations that will mirror those mentioned in the story. After reading the story, you may want to begin a chart with "Our Best Thinking" and find out what your students believe about where the water went and how Vashti might have gotten the better of Juana. A rainy day when there are puddles would be the best possible time to begin this study, but unless you have a very cooperative custodian who would not mind hosing down a section of the playground, you may have to start on a day without rain.

If the children speculate that Vashti picked a puddle that was spread out and shallow while Juana took a small, deep puddle, ask the children if they can find a way to test this in your classroom. You may have to help them see that you are simulating the outdoor version of the story so that it can be tested. A helpful question may be: "Does the size of the opening of a container make a difference in how fast the water in it evaporates?" You may want to introduce them to the term *surface area* even though they may have only a cursory understanding of what *area* means. One way is to show how some surfaces can be covered by more or fewer pieces of paper. For example, a pie plate presents almost a full notebook-size paper's area to the air while a cup presents only one-eighth of a piece of paper. The students will probably notice that in the pie plate, the water is not as deep as it is in the cup but is spread out over a greater area.

Usually the children will want to use an amount of water (e.g., a cup) and put equal amounts of water into a container with a small opening and into a large shallow baking dish. It is a good idea to try this yourself first to see how much time it takes to collect data so you know when to begin the investigation. You will want the investigation to end during the school day, so it must be started early enough. If it is done in the warm weather, placing the container on a windowsill next to an open window will speed up the process, and in cooler weather, putting them on the radiator will also work. Other questions may arise such as

- Does a shallow puddle always evaporate faster than a deeper one?
- How do three or four containers of increasing size affect the evaporation rate? (or words to that effect)
- Does heat speed up the rate of evaporation?
- What is the stuff left behind after evaporation?
- Can we use a liquid that will leave nothing behind?
- What would happen if we dissolved a lot of salt or sugar in the water and then let it evaporate?
- Does a lot of stuff in the water affect the time it takes for the water to evaporate?

Other questions will surely arise and can be tested. As you can see, this activity may take you into the realm of dissolved materials in liquids and their recovery via evaporation. If you wish, this can take you into even more investigations.

CHapter 13

USING THE STORY WITH GRADES 5–8

Working with older children on this topic will be somewhat different since most of your students will at least give verbal agreement that water vapor is present in the air. This does not, however, mean that they actually believe it. You can find out by giving the probe "Wet Jeans" from Page Keeley's book *Uncovering Student Ideas in Science, Volume 1* (Keeley, Eberle, and Farrin 2005). This probe asks students to choose from seven options as to where the water went when blue jeans were hung out to dry. Then they have to provide an explanation. You may be surprised that students will pick the right answer, but when asked to explain how they know, they will be hard-pressed to provide you with a plausible reason for their choice. Some may even believe that water has to be boiled before it can evaporate. This story can help them connect some of their past experiences and then lead them into questions about how Vashti was able to make her predictions about the puddles. If your students understand the scientific model of the particulate nature of matter, they will have little trouble with this idea.

You could treat it like a mystery story: "What did Vashti know that Juana did not?" or "Do you think you could pick out two puddles in the school yard that would re-create the behavior of the puddles in the story?" or "Do you think you could create an investigation that would show that water in different containers evaporates at different rates?" or even, "Does the size and shape of a container affect the rate of evaporation?"

You can also borrow questions from the K–4 section, but I suspect that your students will generate these and many more as the investigations go on.

RELATED BOOKS AND NSTA JOURNAL ARTICLES

Driver, R., A. Squires, P. Rushworth, and V. Wood-Robinson. 1994. *Making sense of secondary science: Research into children's ideas.* London and New York: Routledge Falmer.

Keeley, P. 2005. *Science curriculum topic study: Bridging the gap between standards and practice.* Thousand Oaks, CA: Corwin Press.

Keeley, P., F. Eberle, and C. Dorsey. 2008. *Uncovering student ideas in science: Another 25 formative assessment probes, volume 3.* Arlington, VA: NSTA Press.

Keeley, P., F. Eberle, and J. Tugel. 2007. *Uncovering student ideas in science: 25 more formative assessment probes, volume 2.* Arlington, VA: NSTA Press.

Keeley, P., and J. Tugel. 2009. *Uncovering student ideas in science: 25 new formative assessment probes, volume 4.* Arlington, VA: NSTA Press.

Konicek-Moran, R. 2008. *Everyday science mysteries.* Arlington, VA: NSTA Press.

Konicek-Moran, R. 2009. *More everyday science mysteries.* Arlington, VA: NSTA Press.

Nelson, G. 2004. What is gravity? *Science and Children* 41 (1): 22–23.

references

American Association for the Advancement of Science (AAAS).1993. *Benchmarks for science literacy.* New York: Oxford University Press.

Keeley, P., F. Eberle, and L. Farrin. 2005. *Uncovering student ideas in science: 25 formative assessment probes, vol. 1.* Arlington, VA: NSTA Press.

National Research Council (NRC). 1996. *National science education standards.* Washington, DC: National Academies Press.

CHAPTER 14

THE LITTLE TENT THAT CRIED

Splash! Right in the left eye.

Rani looked up into the darkness inside the tent that was her camping home for the night. Splash!! Right smack in the right eye this time.

"Okay, who's the wise guy?"

Splash!! Right in the middle of the forehead.

"Okay, that's it! Somebody is in trouble and their squirt gun is toast!"

Rani turned on her flashlight only to find the tent tightly zipped up and her tent partner, Annie sleeping soundly. At least, she was pretending to sleep.

"Annie, wake up!" yelled Rani as she shook her friend.

"Wha-, wha-, what's going on?… Why are you waking me up, Rani?" said a sleepy Annie.

"You know what's going on, Annie," said Rani angrily.

"Why is water dripping down your face, Rani? You look like you were in a shower."

"Exactly," spat Rani, "and I feel like I was in a shower too!"

Splat! Right on the pillow behind her.

And now Rani felt a little sheepish. She was looking right at Annie and yet the water was still hitting her bed. She shone the flashlight up on the tent top and there it was, a drop of water waiting to fall on her bed again.

"What do I have to do, sleep under an umbrella?"

"What are you raving about, Rani? It's the middle of the night!" And then Annie looked up at the tent top illuminated by Rani's flashlight beam.

"Oh, no! We have a leaky tent and it must be raining. But at least it's only on one side of the tent. Goodnight, Rani,"

"Oh no you don't, Annie. We share this tent, and if I get wet, you get wet."

"No way! I'm too tired to argue but if you want to slide over to my side, go ahead."

Rani opened the flap on the tent and looked out. The moon and stars were bright, it was cool—but there certainly was no rain.

"There goes that theory," said Rani and snuggled over as far as she could get toward the other side of the tent.

About an hour later: splat! Right in her right ear. This time she was too tired to care and slept the rest of the night.

The next morning, the campers awoke to another hot and muggy day. It had been in the 90s for a week now and it felt like they were swimming in hot air. Rani's pillow was soaking wet and there was plenty of moisture on Annie's pillow as well. Rani had to find out what was going on. It happened that everybody had damp spots in their tents as well. All of their tents couldn't have been leaking and anyway there had been no rain all night. The grass was wet and the leaves on the trees were wet and the inside of all of their tents were beaded with water droplets.

Penny, their counselor, was getting the morning fire started when the girls approached her and told her the story of their wet night.

"That's very interesting," said Penny. "I'll bet you are wondering where the water came from. Do you have any ideas? There has been a lot of humidity lately— you know, a lot of moisture in the air. Maybe it came from there."

Rani and Annie looked at each other. "I certainly didn't feel any water in the air and I've been walking around in it for most of the week," said Rani.

"I really think our tents are leaky," said Annie.

"Everybody's?" asked Penny.

"Well, that is strange, but where else could the water come from and get inside our tents if there wasn't a hole in the tent?"

"Maybe it came from our breath. You know, like when you breathe on a window, it gets cloudy," said Tom, who was standing nearby.

"Yeah, well maybe so, but why did it collect on the tent ceiling and rain on us?" said Annie, unconvinced.

"That sounds like a lot of magic!" muttered Rani. "Invisible water from the air or our breath suddenly turning to rain inside our tents. I think it's time for a morning swim. At least I can see that water without using my wand!"

NATIONAL SCIENCE TEACHERS ASSOCIATION

PUrPOSe

Rani and Annie experienced the water cycle firsthand or perhaps, "first face" would describe it better! This actually happened to me during a camping trip in Everglades National Park years ago. The humidity was fierce but the air cooled down overnight and I awoke to a wet face and wet pillow. I was not sure whether it was the high humidity of the subtropics, or my breath, or both, but I knew one thing for sure: My pillow and I were wet! This story is designed to help the students see the water cycle in a natural situation rather than in the usual highly stylized manner. The water in their breath or in the air inside their tent in vapor form condensing on the cooler tent surface and returning to liquid form "raining" down upon their bodies is something to which the students might be able to relate directly.

reLaTeD CONCePTS

- Evaporation
- Temperature
- Relative humidity
- Conservation of matter
- Condensation
- Humidity
- Cycle and energy

DON'T Be SUrPriSeD

Your students will provide you with many interesting opinions about where the water in the tent came from. Very few children in elementary and middle schools believe that there is actually enough water in the atmosphere to cause this phenomenon. Most of the teachers with whom I have discussed this will tell stories of how difficult it is to get the children to believe that the water that collects on the outside of a cold drink comes from the air around it, but would rather believe that the water penetrated the glass from the inside. In the story, Rani is obviously unconvinced that there is water in the air surrounding her and therefore would be resistant to the idea that it could be responsible for the "rain" in her tent. Rani is typical of most children and some adults in this respect. The moisture in breath is easier to believe since we see evidence of this each time we breathe on a mirror or a glass in the fog that is created. Yet, we know that in such instances as the "sweaty" glass of cold drink, the water must come from the air and that it is indeed present. Convincing children is a different matter.

One factor that has added to the confusion in understanding the water cycle has been the traditional representation of the land, water, clouds, and rain in diagrams in many texts that show water going from ponds and lakes, directly up to the clouds and then raining down upon the source again. This oversimplification can cause children and adults alike to believe that the water cycle is a consistent, never-ending transportation from land to cloud to land again. Given the fact that most of our planet's fresh water is tied up in underground reservoirs, ice and in

oceans, lakes, and streams, water that evaporates spends most of its time in one of these places or else remains in the atmosphere in our general vicinity as atmospheric moisture. Water in the oceans has been locked in place for centuries and the same is true for the glaciers and polar caps. The important concepts here are evaporation, condensation, and the conservation of matter despite the physical change in the state of the water in question. A common misconception is that water no longer exists when it evaporates.

However, we must not ignore the part of the water cycle that brings us clouds, precipitation, and the replenishing of the water on our planet, which we need so badly.

CONTENT BACKGROUND

You may have already experienced a situation like this by finding droplets of water on your ceiling or walls during a particularly humid weather episode. Or you may have noticed fogging windows during a cold spell when you were boiling something on the stove. Another example would be the fogging of the inside of your auto windows on cold days. Contrary to pictorial views of the water cycle in texts, it happens in everyday situations without rain clouds and lakes. You see a form of the water cycle when you, as Tom said in the story, breathe on a cold window and form a fine mist of water on the window. How many of us remember doing that on car trips when we were children and then writing our names (or other words) on the windows, much to our parents' chagrin? Our warm breath containing water vapor loses energy to the cold window and the gas changes into liquid. The gas, warmed by the energy from our bodies, was transformed when it touched the colder window, thus causing what scientists call a phase change. In this case it is called condensation.

Water seems to be a magical substance that falls from the sky, runs from our faucets, and seemingly disappears from wet clothes and puddles. In order to understand the water cycle, one must understand that water molecules can exist suspended in the air, that they get there through the process of evaporation or the escape of these molecules from the surface of water and that the escaped molecules (vapor) can change back into liquid water again (condensation). All of these changes require the give and take of energy. It takes energy transferred to the water molecules in a pair of jeans hanging on the line to change the water molecules to vapor. It requires a release of energy by these molecules in vapor form to become water molecules again. Both the water molecules and the energy involved are conserved; in other words, neither the mass of the water molecules or the amount of energy in the transfer changed.

First of all, water is a liquid, which means that its molecules are in motion and more loosely held, so that they can roll over each other and therefore fill a container or spill out of that container when the liquid is poured. Sometimes the energy in the motion of a molecule is great enough that it can escape the rest of its

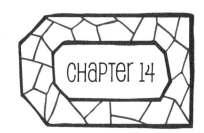

neighbors (evaporate) and suspend itself in the air. Here it joins other molecules of water in the air, which bounce off each other and thereby are able to move up into the atmosphere. As it leaves its fellow molecules to evaporate, a bit of energy was required to break it loose from its surroundings so that the liquid it has just left, which supplied the energy, is a tiny, tiny bit cooler than it was before the escape. You may have noticed this when water evaporates off of your skin when you are wet and you feel cooler since the heat from your body has supplied the energy that allows the water to evaporate. This explains the chill and the "goose bumps" on your skin when you get out of the pool.

Second, these molecules of water can change back again into liquid if they lose the energy to another source. On cool ground, high in the sky, on your cool car window, or in cool air, the vapor molecule loses its extra heat energy and returns to a liquid form. If the vapor touches the cooler surface of a tent, it reverts to water, and as the tiny droplets adhere to each other the droplets become large enough that they eventually fall on your face or pillow.

I do not intend to downplay the water cycle that involves large areas of the globe. These are the cycles depicted in most textbooks showing water evaporating from lakes or oceans, rising up to the sky, forming clouds, and raining the original water back down to earth. The basic phenomenon of the water cycle is absolutely essential to the planet. But as depicted, it gives most students an oversimplified view of the ways in which water is recycled. Students may believe from these drawings that water from puddles in Chicago evaporate around noon, go immediately into the sky, form clouds, and rain the same water later that afternoon in Detroit, from which it evaporates again during the afternoon from Lake Erie. Actually that could happen, but in reality water may spend centuries or longer in oceans and we know that some of the ice in the glaciers might be 10,000 years old. When glacial ice melts, the meltwater, which is full of glacial debris, often goes directly to the bottom of the ocean to remain there for eons before it can rise to the surface and have an opportunity to evaporate. In some cases, the glacial meltwater will mix with surface water but still remains in a huge reservoir of water.

I am particularly fond of a simulation game called "The Incredible Journey," found in the teacher's edition of Project WET (1995). It is a game in which the students act out the journey of water droplets in the cycle, moving from place to place, but often ending up in seemingly endless lines at the ocean or ice cap locations, showing that the cycle is not the idealistic cyclical occurrence that text drawings tend to suggest. You can find this activity on their website at *www.projectwet.org/activities.htm*. Click on "The Incredible Journey." The water cycle is dependent upon numerous conditions, which are the basis for the story and I hope the discussions and investigations that follow.

related Ideas From the National Science Education Standards (NRC 1996)

K–4: Properties of Objects and Materials

- Materials can exist in different states: solid, liquid, and gas. Some common materials, such as water, can be changed from one state to another by heating or cooling.

5–8: Structure of the Earth System

- Water, which covers the majority of the Earth's surface, circulates through the crust, oceans, and the atmosphere in what is known as the "water cycle." Water evaporates from the Earth's surface, rises and cools as it moves to higher elevations, condenses as rain or snow, and falls to the surface where it collects in lakes, oceans, soil, and in rocks underground.

related Ideas From Benchmarks For Science Literacy (aaas 1993)

K–2: The Earth

- Water left in an open container disappears, but water in a closed container does not disappear.

3–5: The Earth

- When liquid water disappears, it turns into a gas (vapor) in the air and can reappear as a liquid when cooled, or as a solid if cooled below the freezing point of water. Clouds and fog are made of tiny droplets of water.

6–8: The Earth

- The cycling of water in and out of the atmosphere plays an important role in determining climatic patterns. Water evaporates from the surface of the earth, rises and cools, condenses into rain or snow, and falls again to the surface. The water falling on land collects in rivers and lakes, soil, and porous layers of rock, and much of it flows back into the oceans.

Using the Story With Grades K–4

If you can obtain a copy, you might want to begin with using the probe "Wet Jeans" from *Uncovering Student Ideas in Science, Volume 1* (Keeley, Eberle, and Farrin 2005). This could also be used as a pre- and postassessment.

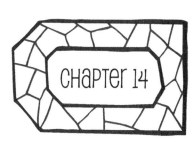

This chapter's story is geared toward students who are at least eight or nine years old. Five- and six-year-old children may enjoy the story and provide you with interesting responses but will very likely be too young to respond to the intent of the story. In fact the Standards and Benchmarks both expect that at the K–2 age level, students should be focusing on the observations of water "disappearing" (evaporating) from puddles and dishes, and so on. I have a problem with the word *disappearing*, because it often means, "no longer existing" to children. Another definition states that disappearing means, no longer in view, which is more accurate in this case. However it might be best to use an analogy such as: A ball rolling under a chair seems to disappear but it is still there although hidden from view. For the third and fourth graders the Standards and Benchmarks both agree that the concept of water changing to vapor and back again is not too difficult. For the younger students, activities inquiring into the conditions that accelerate or hinder the evaporation of water are useful. You might ask them what things help or hinder the water from escaping from a dish of water. A good question to begin this inquiry might be, "How many ways can we think of to make water evaporate faster?"

You will have to help them identify variables such as the surface area of the dish (shallow water in large dish) or depth of water, (deeper water in a small dish or glass), and keeping the amount of water constant in all tests, placing the dishes or glasses in the same spot, and so on. If they look at the process as a race, they can predict from their own experiences and concentrate on making the race "fair." They will find that shallow water and large surface area cause the most rapid evaporation and the idea of a large, shallow surface causing faster evaporation should not be lost on them.

When using this story with third or fourth graders, you can find suggestions below in the section on using the story with grades 5–8 and modifying them accordingly.

USING THE STORY WITH GRADES 5-8

As with all stories, after the reading you should ask the students what they know about the problems posed in the story. Write their comments on a large sheet of paper labeled "Our Best Thinking So Far." When these statements are turned into questions, the students may begin to pose hypotheses to test. All of these steps should be recorded in their science notebooks. If you are comfortable with having several experiments going on at the same time, small groups of students can choose a hypothesis to test and then go about designing the investigation. It usually makes a lot of sense to have these design groups report to the class and ask for suggestions. That way the entire class is involved to some degree in each investigation. Usually students seem to want to reproduce the situation featured in the story. Small tents can be constructed from coat hangers and fabric such as oil cloth, canvas, or rip-stop nylon. It will be necessary to cool off the tent surface; this can be done with a plastic bag of ice cubes hung on the tent surface. Students may want to breathe into the tent to simulate the breathing by the sleeping girls.

Some children like to place a saucer of warm water in the tent and let it evaporate. Soon, under the area that has been cooled, water droplets will form, reproducing the situation in the story. In the discussion that follows, you can introduce students to the terms *evaporation* and *condensation*, which will now have a real-world connection.

From the chart you created at the beginning of the story follow-up you will probably have children who will tell you about drink glasses that were coated with water or other experiences with condensation or evaporation. These statements need to be tested experimentally and shared with the class by summaries from their science notebooks. After playing the game "The Incredible Journey" from the *Project Wet* guide, the topic of global water cycles can be visited and at this point, the diagrams of the water cycle can be discussed with experience and knowledge about the concepts involved, evaporation and condensation. Bringing in the points about energy gain and loss can be done with students who have the maturity to discuss such things. At this point however, it is enough that they have had the firsthand experience with the water cycle system and with its component parts.

related Books and NSTa Journal articles

Driver, R., A. Squires, P. Rushworth, and V. Wood-Robinson. 1994. *Making sense of secondary science: Research into children's ideas.* London and New York: Routledge Falmer.

Hand, R. 2006. Evaporation is cool. *Science Scope* (May): 12–13.

Keeley, P. 2005. *Science curriculum topic study: Bridging the gap between standards and practice.* Thousand Oaks, CA: Corwin Press.

Keeley, P., F. Eberle, and L. Farrin. 2005. *Uncovering student ideas in science: 25 formative assessment probes*, volume 1. Arlington, VA: NSTA Press.

Keeley, P., F. Eberle, and J. Tugel. 2007. *Uncovering student ideas in science: 25 more formative assessment probes*, volume 2. Arlington, VA: NSTA Press.

references

American Association for the Advancement of Science (AAAS). 1993. *Benchmarks for science literacy.* New York: Oxford University Press.

Keeley, P., F. Eberle, and L. Farrin. 2005. *Uncovering student ideas in science: 25 formative assessment probes*, volume 1. Arlington, VA: NSTA Press.

National Research Council (NRC). 1996. *National science education standards.* Washington, DC: National Academies Press.

Project WET, Curriculum and activities guide. 1995. The amazing journey. Bozeman, MT: Water conservation Council for Environmental Education, 161–165.

CHAPTER 15

WHERE ARE THE ACORNS?

Cheeks looked out from her nest of leaves, high in the oak tree above the Anderson family's backyard. It was early morning and the fog lay like a cotton quilt on the valley. Cheeks stretched her beautiful grey, furry body and looked about the nest. She felt the warm August morning air, fluffed up her big grey bushy tail and shook it. Cheeks was named by the Andersons since she always seemed to have her cheeks full of acorns as she wandered and scurried about the yard.

"I have work to do today!" she thought and imagined the fat acorns to be gathered and stored for the coming of the cold times.

Now the tough part for Cheeks was not gathering the fruits of the Oak trees. There were plenty of trees and more than enough acorns for all of the grey squirrels who lived around the yard. No, the problem was finding them later on when the air was cold and the white stuff might be covering the lawn. Cheeks had a very good smeller and could sometimes smell the acorns she had buried earlier. But not always. She needed a way to remember where she had dug the holes and buried the acorns. Cheeks also had a very small memory and the yard was very big. Remembering all of these holes she had dug was too much for her little brain.

The Sun had by now risen in the east and Cheeks scurried down the tree to begin gathering and eating. She also had to make herself fat so that she would be warm and not hungry on long cold days and nights when there might be little to eat.

"What to do... what to do?" she thought as she wiggled and waved her tail. Then she saw it! A dark patch on the lawn. It was where the Sun did not shine. It had a shape and two ends. One end started where the tree trunk met the ground. The other end was lying on the ground a little ways from the trunk. "I know," she thought. "I'll bury my acorn out here in the yard, at the end of the dark shape and in the cold times, I'll just come back here and dig it up! Brilliant Cheeks," she thought to herself and began to gather and dig.

On the next day she tried another dark shape and did the same thing. Then she ran around for weeks and gathered acorns to put in the ground. She was set for the cold times for sure!

Months passed and the white stuff covered the ground and trees. Cheeks spent more time curled up in her home in the tree. Then one bright crisp morning, just as the Sun was lighting the sky, she looked down and saw the dark spots, brightly dark against the white ground. Suddenly she had a great appetite for a nice juicy acorn. "Oh yes," she thought. "It is time to get some of the those acorns I buried at the tip of the dark shapes."

She scampered down the tree and raced across the yard to the tip of the dark shape. As she ran, she tossed little clumps of white stuff into the air and they floated back onto the ground. "I'm so smart," she thought to herself. "I know just where the acorns are." She did seem to feel that she was a bit closer to the edge of the woods than she remembered, but her memory was small and she ignored the feelings. Then she reached the end of the dark shape and began to dig and dig and dig!

And she dug and she dug and she dug! Nothing! "Maybe I buried them a bit deeper," she thought, a bit out of breath. So she dug deeper and deeper and still, nothing. She tried digging at the tip of another of the dark shapes and again found nothing. "But I know I put them here," she cried. "Where could they be?" She was angry and confused. Did other squirrels dig them up? That was not fair. Did they just disappear? What about the dark shapes?

How can she find the acorns? Where in the world are they? Can you help her find the place where she buried them? Please help, because she is getting very hungry!

NATIONAL SCIENCE TEACHERS ASSOCIATION

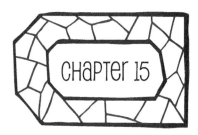

PURPOSE

The main purpose of the "Cheeks" story is to get the children to learn something about the behavior of shadows cast by objects in sunlight. Although the story takes liberties with the "thoughts and projections" of Cheeks, one can take it as merely a motivational story. Some may be concerned with anthropomorphism but children read stories every day about animals that talk and have emotions. To leave these aspects out of the story would remove the "hook" that connects the students to the story characters. We believe that the teacher can make sure that children do not use these liberties to further misconceptions about the animals involved.

Primarily, the story addresses the motion of the Sun in the sky throughout the seasons, or what is called *daytime astronomy*. It is unfortunate that both of these conceptual areas are often relegated to students merely reading about the Sun's path through the sky during the seasons. Books and diagrams without the benefit of observation have become the main entryway for students to learn about Earth-Sun relationships. This is not necessary since measurable data about these motions are readily available to all students who live in places where the Sun shines fairly consistently.

RELATED CONCEPTS

- Rotation
- Revolution
- Earth-Sun relationships
- Measurement
- Periodic motion

- Axis
- Seasons
- Shadows
- Time
- Patterns

DON'T BE SURPRISED

Your students may well feel that Cheeks is a victim of thievery if they have little experience with shadow lengths changing, either daily or seasonally. Others may have little understanding of shadows at all! It is not uncommon for students to believe that shadows bounce off objects rather than be caused by their blocking light. I have personally witnessed 7- to 10-year-olds who had never played with shadows in any way. At the same time other students will immediately suspect the reason for Cheeks' error in judgement. Playing games such as shadow tag or messing about with shadows may prove useful.

A common misconception held by children and adults alike is that summer is warmer in the northern hemisphere because Earth is closer to the Sun. Actually, due to Earth's slightly elliptical orbit and most surprising to most people, Earth is farther from the Sun during the time it is summer in the northern hemisphere. Another common belief is that the Sun is directly overhead at noon and there is no shadow cast at noon. Some students may also believe that shadows do not change in length at all, either daily or seasonally.

I cannot stress enough how important it is for you to collect some data by doing some shadow measurements yourself before trying this with your students. It will prepare you for the potential problems they may encounter and will give you some insight into what kinds of data they will bring into the classroom for analysis.

CONTENT BACKGROUND

The easiest way to begin is to find a place in your immediate area that is sunny most of the day. Place a dowel or pencil in the ground in the middle of a piece of paper that is attached to the ground by toothpicks so that the wind will not move it around. This stick is known as a Gnomon or shadow stick. (See Figure 15.1.) Make sure you put it in a level spot and then check the shadows cast by the gnomen every hour or so. You want to mark the shadow with a line that outlines the shadow. Mark the time as well as the outline of the shadow. After a few hours, you will notice that the shadows cast by the gnomon move clockwise around the paper and as the day approaches midday,

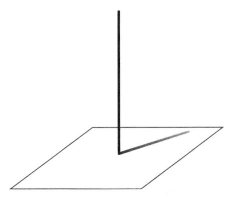

Figure 15.1. Gnomon

the shadow becomes shorter. It will be at its shortest at midday, then begin to lengthen again. You will also notice that the Sun's apparent motion in the sky will correspond to the shadow cast in an opposite direction. In other words, the shadow will be pointing away from the Sun as it appears to move from east to west across the sky. This should give you a clue as to what to expect on a daily basis. You noticed that length and direction of the shadow change over the period of the day you observed. It should also have become apparent that as the Sun rose higher in the sky on its daily path, the shadow became shorter and that at the beginning of the day and after midday, the Sun moved lower in the sky and correspondently the shadow lengthened again (low Sun = long shadow, high Sun = short shadow). Your second observation should be that the shadow always pointed away from the Sun so that it moved from west to east as the Sun moved across the sky from east to west.

If you were able to do this for a whole school year, starting in the autumn (and the following observations are correct only if you start in autumn), you would notice that the shadows would not fall in the same place on the paper from one day to the next. They would fall a bit further counterclockwise to previous shadows. Using what you found out about the relationship between the Sun and the direction of the shadow, you would surmise that the Sun had shifted its position at any given time a little to the southeast. In the northern hemisphere this would mean that as autumn proceeded and winter approached,

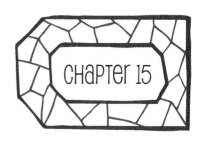
the Sun would rise later each day and would rise a bit more to the southeast than the day before. The Sun is spending less time in the daytime sky until December 21st (sometimes December 22nd) when winter officially arrives, also known as the "shortest day of the year." You should purchase a copy of the *Old Farmer's Almanac* at your local supermarket, bookstore, hardware store, or garden store. Specific tables give times for sunrises and sunsets for each day of the year. There are editions for each section of the country. If you do not have access to this book, the local newspapers also have an almanac that will provide you with astronomical times. You can also access the *Old Farmer's Almanac* online at *www.almanac.com.*

After December 21st, the "winter solstice," you will notice the opposite trend taking place. The Sun will be "rising" each day more toward the north and your shadows will correspondingly shift to this motion as you record them. The shadows will become shorter at any given time when compared to the shadows taken as winter approached. Your daily shadows will move even more to the south as summer approaches and the length of the day increases substantially. Unfortunately, school will probably end before the summer solstice on June 21st (sometimes June 20th), when your midday shadow would measure the shortest of the year. Perhaps, if you are fortunate, you students will become interested enough to continue gathering data through the summer so that they can witness the entire cycle. From June 21st on, your shadow measurements will begin to lengthen again and the cycle will repeat as the next autumn and winter approaches. You would have had to make a one-hour adjustment in your data collection for Daylight Saving Time in March. You would notice that on the day the clocks were changed to one hour ahead, the shadows you record would be an hour behind. You would therefore have to take your readings adjusted to Sun time rather than clock time. It is important to notice that Sun time is "real" time and that changing clocks does not alter the astronomical movement of the celestial bodies. I am reminded of the joke about the gardener who opposed daylight saving time because he thought that the extra hour of sunlight would be bad for his crops.

Your observations will eventually lead you to conclude, by means of your records of the motions of the Sun and the records of the corresponding shadows, that the Sun prescribes a predictable path in the sky and that each year this cycle continues. Further study may also lead you to the evidence that this motion is tied to the reasons for our northern and southern hemispheric seasons. This periodic motion is but one of many in the universe. In Chapter 4 ("Moon Tricks"), you were introduced to the periodic motion of the Moon and its phase cycles. Each year you witness the periodical motion of the Earth and Sun, which causes the seasons, and each day, the rotation of the Earth to cause day and night. In the story "Grandfather's Clock" (*Everyday Physical Science Mysteries* 2013) you witness the periodic motion of the pendulum, which has a cycle so dependable that you can use it to keep time. It is no wonder that the big idea of periodic motion is so important as a unifying concept in science. It is also evident that science seeks patterns that eventually lead to predictions, which lead to a better understanding of our universe.

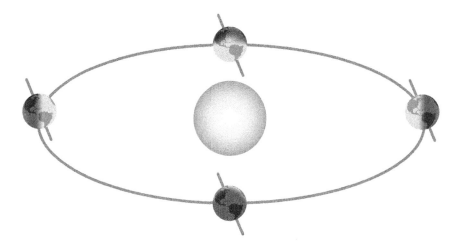

Each year the Earth makes one revolution around the Sun. In the northern and southern hemisphere a tilt of 23½ degrees accounts for the seasons. Since the Earth's tilt always points in one direction, relative to the Earth's orbit there are times when the northern hemisphere is pointed more directly toward the Sun and therefore receives more heat from its direct rays.

This occurs at the summer solstice in June, the beginning of summer; days are long and nights are short. At the other extreme of its orbit, one half year later, the northern hemisphere is pointed away from the direct rays of the Sun at the winter solstice and days are short and nights are long. It is exactly the opposite in the southern hemisphere where the seasons come at opposite times of the year compared to the northern hemisphere. In between these two extremes, the Earth is in transition to either spring or fall and milder temperatures are common since the direct rays of the Sun fall more evenly above and below the equator. Days and nights are more equal in length. The tilt of the Earth is the main cause of seasons and of the differences in the Sun's position in the sky during the year and therefore the difference in shadow patterns. One helpful fact is that the farther north or south of the equator you are, the greater the differences there are in day-night hour lengths and seasonal shadow lengths.

related Ideas From the National Science Education Standards (NRC 1996)

K–4: Objects in the Sky
- The Sun, Moon, stars, clouds, birds, and airplanes all have properties, locations, and movements that can be observed.

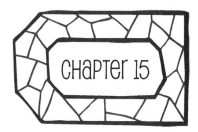

K–4: *Changes in Earth and Sky*

- Objects in the sky have patterns of movement. The Sun for example appears to move across the sky in the same way every day. But its path changes slowly over the seasons.

5–8: *Earth in the Solar System*

- Most objects in the solar system are in regular and predictable motion. Those motions explain such phenomena as the day, the year, phases of the moon, and eclipses.

related Ideas From Benchmarks For Science Literacy (aaas 1993)

K–2: *The Universe*

- The Sun can be seen only in the daytime but the moon can be seen sometimes at night and sometimes during the day. The Sun, Moon, and stars all appear to move slowly across the sky.

3–5: *The Earth*

- Like all planets and stars, the Earth is approximately spherical in shape. The rotation of the Earth on its axis every 24 hours produces the day/night cycle. To people on Earth, this turning of the planet makes it seem as though the Sun, Moon, planets, and stars are orbiting the Earth once a day.

6–8: *The Earth*

- Because the Earth turns daily on an axis that is tilted relative to the plane of Earth's yearly orbit around the Sun, sunlight falls more intensely on different parts of Earth during the year.

USING THE STORY WITH Grades K–4 and Grades 5–8

Please revisit the introduction to this book for the case studies about how two classroom teachers used this story. It will help you to see the overall picture of the process. Since the case study is available I will combine the grade level suggestions in one section. There are also a great many common elements for both levels, so to avoid repetition, please read ideas for both grade levels and choose the ideas most appropriate for your class.

Cheeks, of course, fell victim to the misconception that shadows caused by the blockage of sunlight do not move or change shape during the course of the day or over the seasons. We have discovered that most children, from first grade on, immediately suspect that the shadow's position in the story had changed. The students do not often realize that the Sun's apparent motion is the cause of the changes in the shadow, although one or more students may suggest that this is the case. Students' experience is such that when pressed to explain something like Cheeks' problem, they begin to recall their simple knowledge of shadows and apply it to explain why the acorns were "lost." Once the older children have agreed that shadows do indeed change from day to day, they usually wonder in what ways they change and by how much. As seen in the case study in the introduction, younger children may be satisfied with focusing on what happens to the length of shadows during the course of a school year.

The story might raise some other hypotheses in the minds of the children about what happened to Cheeks' acorns. A few children may carry on with Cheeks' initial suspicion and make up ideas about thieving squirrels or chipmunks. This is a productive entrance to the literacy possibilities open to the teacher. These ideas can be encouraged by suggesting they write some creative narratives about Cheeks, who might be imagined to star in a multitude of adventures in the backyard. These are great entries in the science notebooks. Encourage the children to back up their stories with some facts they can uncover about the behavior of squirrels. This is an excellent door into the secondhand inquiry using textual or internet information mentioned in Chapter 3. For example, when children learn something about what a squirrel's nest looks like and how it is built, stories can emerge from or around these bits of knowledge. I have had children suggest that the acorns sprouted after they were hidden. This could lead into another investigation in the biological area about seeds to see if all acorns germinate immediately when planted. If the teacher feels comfortable in having two or more investigations going on at once, it can lead to an exciting set of concurrent experiments. These can be sideline excursions for some children even though the main thrust is aimed at the discoveries they will be making about the Sun's movement and the shadows that mark it.

With both older and younger children the facilitator should help the children realize that they need to explore what happens to shadows out of doors during the day. Older students are usually more interested in what happens to shadows over the seasons. With both student groups, it might be a good idea to let them play with flashlights and objects to study shadows and their relationship to light sources. Productive questions to ask might include the following:

(1) Can you make the shadow change its length? How did you do that?
(2) Can you make a shadow that is longer or shorter than the object? How do you do that?
(3) Can you make the shadow move around the table (desk) in different directions? How do you do that?

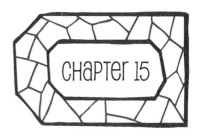

These kinds of observations also belong in their science notebooks along with questions that arise from their explorations.

Once they have had practice with making shadows, the ideas for Cheeks' problem may become more obvious to them. Younger children may come up with knowledge statements recorded on the "Our best thinking" sheet, such as the following:

- Outdoor shadows get longer as the day goes on.
- Outdoor shadows get shorter as the day goes on.
- Outdoor shadows change all the time during the day.
- Outdoor shadows point in different directions during the day.

Older students may have other misconceptions that they will share. These might include:

- At noontime there will be no shadow because the Sun is directly overhead.
- Noontime and midday occur at the same time.

As suggested previously, these knowledge statements can be changed to questions so the knowledge statements become productive questions, e.g. "Do outdoor shadows get longer as the day does on?" "Will there be a shadow at noontime?" These are obviously testable questions and can be changed back to hypotheses and tested. This may seem like an unnecessary step to go from statement to question back to statement, but I believe it helps the children to see that hypotheses come from questions and that all knowledge should be open to question. It should also reveal that a hypothesis is a statement, not a question. Children should also be asked to give some reasons for their hypotheses. They need to learn that hypotheses are not merely wild guesses.

Once the children have given their opinion of why the acorns were missing, the adult can write these down on the "Our Best Thinking for Now" chart. I suggest that this list of theories or guesses about the motion of shadows be written on large pieces of paper and displayed in a public place. It should become a record of their "best thinking so far," and be modified as new ideas are incorporated into their thinking as a result of their activities. In this way they can revisit old ideas and see how they match with their new thinking. With new vocabulary, it is also a great help to ELL students. It helps them to remember where they have been as well as helping them see where they are going and that changing one's mind due to evidence is not a weakness. The hardest part of the adult role is helping the children learn how to look without telling them what to see.

The facilitator can help the children by asking some questions that can help the children focus on the problem and some solutions. Some of these questions for all students might be:

- What did Cheeks expect would happen to the shadows she used as markers?
- What do you think happens to the length and shape of shadows during one day?
- What can we do to find out what happens to the length and shape of shadows during one day?
- How can we make and keep records of what we find out?

For older students you might also ask:

- Do you think that the shadow Cheeks used changed in ways other than in length?

Once the children have agreed on how they will study the shadows during a day and several days recordings are on display, the next question might be:

- How will we find out what happens to shadows over a longer period of time, such as fall to winter or winter to spring?

The next section on methods will discuss the use of the gnomon to collect data to answer these questions.

This last question and discussion of the data collection methodology will bring up many design problems that you must be prepared to address. The children will probably want to use a tree as a marker since Cheeks' dilemma is based on tree shadows. However, during the winter months when the Sun is low in the sky, tall trees can cast very long shadows that can be interrupted by the school building or areas of the school site that cannot be entered because of brambles or fences, thereby frustrating measurements. Explain to the students that since we do not know if or how these shadows can change, they might choose a shorter tree or object in the center of a wide space, which will allow for all sorts of surprises without making data collecting difficult or impossible. One might be tempted to allow the students to find this out for themselves, but in the case of a study over a long period of time, this error can destroy not only the value of the data but the incentive to continue the study. Another way to prevent such disasters is to have several sites using teams of children at each site. If one site should run into problems, there are always data from the other sites. However, it is important that all teams agree on one method of collecting data so that any comparisons that might be required would be compatible. As a result of the discussion of the Cheeks story the children should be able to begin designing the data collection to answer their particular questions.

One of the time-honored methods for collecting shadow data is the gnomon. A gnomon is a stick that is placed in the ground or into a surface parallel to the ground with the stick perpendicular to the ground and which acts as an unchanging shadow producer. The stick blocks the Sun's rays and casts a shadow onto the ground in the shape of the stick. As the day progresses the shadow will change in two ways. It will become longer or shorter in length as the Sun gets higher or lower in the sky, and it will point in different directions as the Sun moves across the sky from east to west. Depending upon the position of the Sun, the shadow may be shorter, the same as, or longer than the stick. It is important that the surface upon which the shadows are cast is flat, not undulating or sloping. Of course, as long as shadows are measured at this exact spot each time, the shadows for each measurement will be comparable. If however, the placement of the gnomon is changed from time to time, as it probably will be, the level of the surface is very important. Paper can be placed on a board and the stick attached to the board at its center so that the shadows will be cast on the paper and can be recorded with pencil or

felt-pen marker. You must only be careful that the place where the board is placed is level and not sloping.

If the gnomon is attached to a board, the board should be pointed in the same direction each time a recording is taken so the position changes can be noted. By this I mean that the sides of the board should be pointing in the same compass direction each time a reading is taken. If you want a really great class discussion,

Figure 15.2. Tree Gnomon

ask the students if it is important to align the board the same each day. The gnomon stick itself should be about the length of a toothpick so that the long winter shadows in higher latitudes do not go off the paper. You may want to add a triangle to the toothpick so that it looks like a pine tree and the story line is kept intact (see Figure 2).

Younger children can measure the shadow lengths with yarn or string and transfer the yarn to a paper to create a graph. Be sure to mark the dates and times carefully on the graph. You might want to have your children keep records of their measurements in their science notebooks. Children who are able to measure can transfer their measurements to a piece of centimeter graph paper and can compare lengths as well as directions of shadows. If these are placed on transparencies, shadow records from various dates and times

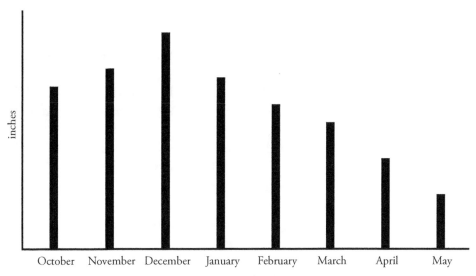

Graph of Shadow Lengths, October to May

can be superimposed on each other and compared for lengths and direction. It helps a great deal if each of the date's shadows are recorded in different colors so that comparisons can be made easier.

The children will find that there are several changes in shadows over short and long time intervals. These main observations can be listed as follows when considering shadows created by the sunlight. These concepts are:

- Shadows change daily and from day to day.
- Shadows always point away from the source of light.
- Shadows are longest in the early mornings and late afternoons.

- Shadows are shortest during the midday hours.
- The shadows change from longer to shorter and back to longer during one day.

Older students may add the following as well:

- Shadows point to the west in the morning and to the east in the afternoon.
- Contrary to expectations, in latitudes higher than 23.5° north or south of the equator there is no time during the day when the Sun is so high that no shadow is cast.
- The shortest shadow is not always at noon. The shortest shadow is cast at midday, which is the midpoint between sunrise and sunset (often called local noon).
- Shadows at any given time change in length and direction as the year progresses.

All of these observations can help the students to understand the motion of the celestial bodies and ultimately the reasons for seasons. It is best to wait until middle school to expect any real understanding of the causes of the seasons. Their spatial relations will have developed by then and their ability to see the spatial relationships of the Sun and Earth will improve.

related Books and NSTa Journal articles

Bogan, D., and D. Wood. 1997. Simulating Sun, Moon, and Earth patterns. *Science Scope* 21 (2): 46–47.

Driver, R., A. Squires, P. Rushworth, and V. Wood-Robinson. 1994. *Making sense of secondary science: Research into children's ideas.* London and New York: Routledge Falmer.

Keeley, P. 2005. *Science curriculum topic study: Bridging the gap between standards and practice.* Thousand Oaks, CA: Corwin Press.

Keeley, P., F. Eberle, and L. Farrin. 2005. *Uncovering student ideas in science: 25 formative assessment probes,* volume 1. Arlington, VA: NSTA Press.

Keeley, P., F. Eberle, and J. Tugel. 2007. *Uncovering student ideas in science: 25 more formative assessment probes,* volume 2. Arlington, VA: NSTA Press.

references

American Association for the Advancement of Science (AAAS). 1993. *Benchmarks for science literacy.* New York: Oxford University Press.

National Research Council (NRC). 1996. *National science education standards.* Washington, DC: National Academies Press.

CHAPTER 16
THE COLDEST TIME

National Park Ranger Rudi was in charge of the program in the Everglades National Park called "Into the Wild." In this program, different families from the Miami, Florida, area go camping for two nights. The families were picked from those who had never been to the Everglades or ever camped outdoors. It was not hard to find families that hadn't done either, but it was difficult to convince them that they should try out the experience. The kids—and even the adults—were afraid of the darkness and the wild animals in the area. Finally Rudi managed to find one family who was willing to go. The family had a mom, dad, and two children—a girl, Sasha, and a boy, Gene.

When they got to the park, Mom said that she hadn't gotten a wink of sleep the night before. She was worried about her children. Would they like the experience? Would they be safe? Would they learn about the Everglades and want to go back?

The family met Ranger Rudi in the parking lot of the visitor center. They did not own any of their own equipment, so Rudi was prepared with binoculars, food, cooking equipment, water bottles, sleeping bags, and tents. It was the dry season in January, but the nights could get cool, so Rudi had nice warm sleeping bags and jackets for everyone since people in Miami were not used to cool weather and did not own much in the way of warm clothes.

After a full day of activities in the park including looking at birds, turtles, fish, and alligators, Rudi helped the family pitch their tents in the campground. There were lots of other visitors around so that the family felt safer surrounded by people. They had a campfire, toasted marshmallows and hot dogs, and soon it was time for bed. It got dark at about 5:30 p.m. and the Sun was due up the next morning at about 6:00 a.m.

"If it gets cold in the middle of the night, can I come and get in with you?" said Sasha, the littlest one, to her parents.

"Sure," said Dad, "As long as it is not really early in the morning."

"Oh, it won't be," said Gene, " 'Cause it will be coldest around midnight."

"I'll bet it will be coldest about three in the morning," said Sasha. "And that *is* early in the morning, real early!"

"Well, when I'm talking about early in the morning, I mean about sunrise," said her father. "Don't come knocking at my tent when I'm just settling down for my pre-coffee sleep."

Rudi broke in and asked, "When do you all think it will be the coldest outdoors tonight, just after midnight, at three in the morning, or at sunrise?"

Sasha said three in the morning, Gene stuck to his prediction of midnight, and Dad was afraid it was going to be right around dawn. Mom was already in her bag and didn't have a clue. She just wanted to get the tent zipped up so no critters could get in.

Rudi said that it might be interesting to find out who was right, but how could they do it?

"Let's sleep on it," her father said, and they did. And the next morning, while they were eating pancakes cooked over a fire, they discussed how they could do it.

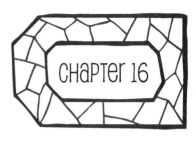

PURPOSE

The purpose of this story is to help students learn about the source of heat energy that warms their planet. Of course that is the Sun, and it only has an effect on the temperature of the Earth when it is shining on a particular spot. Another purpose is to stimulate the students to design a way to find out when, under normal conditions, the temperature in their area is lowest.

RELATED CONCEPTS

- Solar energy
- Heat
- Weather
- Temperature
- Radiational cooling
- Climate

DON'T BE SURPRISED

Your students may be unaware of the radiational cooling that takes place during the night when there are clear skies that allow heat energy to leave the Earth's surface. They may also be of the opinion that the "witching hour" of midnight has some hidden meaning or that sometime in the middle of the night the promise of sunshine will begin to warm the Earth. We have also found that few students realize that even after the Sun appears on the horizon it takes a while before the Sun is high enough in the sky to actually warm the Earth in any particular location.

CONTENT BACKGROUND

Most students and adults would agree that the Sun is the source of heat that warms our planet. We hear enough about radiational cooling from the TV weather reports to know that nights with clear skies allow the heat energy of the Earth to radiate out into space, unimpeded by insulating cloud layers. This simple law of thermodynamics is so much a part of our lives that the idea of heat leaving a warm place and moving to a cooler place is usually not a problem. No, the problem, as shown in the story, is predicting when it is that the heat starts returning. Since the glow of the sunrise signals the appearance of our heat source, it is reasonable to assume that a rise in temperature is expected immediately. But, anyone who has been outdoors at sunrise has felt the cold air and experienced the delay of the eventual warming. It is difficult to understand that more direct rays of the Sun are needed to produce any discernable difference in air temperature.

The Sun's rays have to actually warm up the surface of the Earth, so they must be at an angle that is sufficiently high to make a difference. An analysis of the hourly temperature increases show that the temperature rises slowly in the early morning hours and often does not reach highs for the day until late afternoon.

This means that as the day progresses, and if the Sun continues to shine, the Earth absorbs more heat than it radiates back so that it warms up. This implies that the Earth is continuously receiving heat energy and is also losing it at the same time. Due to the high angle of the Sun in the afternoon hours, the incoming heat exceeds the outgoing heat, sometimes even after sundown, when the Earth may lose heat slowly due to cloud cover.

For those who live in a climate where winter frost leaves a coating on the ground, there is a dramatic source of evidence for radiational heating. On the white frosting of ice crystals on the ground, we can see shadows of trees, posts, and other objects. As the sun rises higher in the sky, we see something I call "frost shadows," because frost begins to melt *except* where there is a shadow that blocks the Sun's rays from striking the ground. So these white "shadows" are caused by the lack of radiation from the Sun. As the Sun rises higher and higher, the white shadows slowly disappear as the angle and direction of the Sun changes. This shows that the Sun's rays are responsible for the warming of the Earth's surface, eventually melting the frost. It also demonstrates, indirectly, that the temperature of the Earth's surface takes some time to rise, which is why the frost hangs around for a while. And it provides evidence that the lower angles of the Sun's rays are not as strong as the higher angles.

This phenomenon also shows that the Sun's position in the sky changes not only in its rise above the horizon, but toward the south and eventually the west. Thus, the white shadows begin to disappear in a clockwise direction as the Sun changes position. The wider the shadow, the easier it is to see this happen and perhaps even measure the Sun's movement, both in direction and speed. Watching the frost shadows disappear from a large tree's shadow takes more time than from a thin pole or sapling. For teachers, the timing of the phenomenon of frost shadows is great. By the time the Sun rises high enough to peek over the roof of the school, the frost is still on the ground. Classes can observe the frost during the opening hours of the school day and have plenty of time for discussion afterward.

During this discussion, someone usually beings up the topic of climate. A student may say, "Climate is what you expect, weather is what you get." Climate, however, is determined by many variables such as altitude, latitude, proximity to water, wind, rainfall, and other conditions that are examined and recorded over a period of at least 30 years. Thus we have deserts, rain forests, cloud forests, tropics, tundras, and so on. For example, south Florida is listed as a subtropical climate, even though other places around the world at the same latitude are deserts. What makes its climate different than the Sonoran, Gobi, or Sahara deserts is that south Florida is surrounded by water and warm currents that combine to provide it with a wet season and a dry season. It is classified as subtropical because the usual definition of *tropical* entails that the latitude is within 23.5° north or south of the equator. South Florida is about 25° north of the equator, so, along with its unusual tropical weather, it can get frost and even extended freezes, like the one just experienced during the winter of 2009–10. These are rare, occurring only once or twice a century, however.

In the context of this story, climate might have a distinct effect upon the temperature range seen in one day. For instance, in a desert, the temperature might

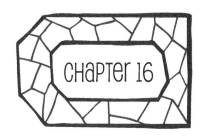

range from freezing (below 32° Fahrenheit) at night to the low hundreds during the day. This would correspond with the situation described in the story with the lowest temperatures being around dawn. But, if this area were surrounded by mountains that block the Sun's rays, it would delay the rise in temperature until the Sun was at an altitude higher than the surrounding mountains.

Radiation or *radiative cooling* is the process whereby the Earth loses heat by emitting long-wave (*infrared*) radiation, which balances the short-wave or *visible* energy from the Sun. The cooling of the Earth is very complex, having to do with transference of heat due to convection (air currents), evaporation, and other things, such as geographical factors. Radiative cooling goes on all the time, but is especially intense under conditions such as nighttime clear skies, light winds, and low humidity.

related ideas from the National Science Education Standards (NRC 1996)

K–4: Objects in the Sky
- The Sun provides the light and heat necessary to maintain the temperature of the Earth.

5–8: Transfer of Energy
- The Sun is a major source of energy for changes on Earth's surface. The Sun loses energy by emitting light. A tiny fraction of that light reaches the Earth, transferring energy from the Sun to the Earth. The Sun's energy arrives as light with a range of wavelengths consisting of visible light, infrared radiation, and ultraviolet radiation.

5–8: Earth in the Solar System
- The Sun is the major source of energy for phenomena on Earth's surface.

related ideas from Benchmarks for Science Literacy (AAAS 1993)

K–2: Energy Transformations
- The Sun warms the land, air, and water.

3–5: The Earth
- The weather is always changing and can be described by measurable quantities such as temperature, wind direction and speed, and precipitation.

6–8: The Earth
- The temperature of a place on Earth's surface tends to rise and fall in a somewhat predictable pattern every day and over the course of a year.

6–8: Energy Transformation
- Heat can be transferred through materials by the collisions of atoms or across space by radiation.

USING THE STORY WITH GRADES K–4

One of the first things that would benefit younger children would be to become familiar with the thermometer—how to read one and the range of a typical thermometer. Many teachers prefer to begin with a thermal strip thermometer, the kind that is used on the forehead to take body temperatures. Prepare several containers of water that are of various temperatures, yet within the comfort and safety zone. Children immerse their hands in the water and then compare the sensation with the reading on the thermometer. The concepts of *cool* and *warm* may then take on a numerical value. They may realize that the thermometer is a better indicator of temperature change than their senses.

To get an idea of the power of the solar energy that comes from the Sun, you may want to purchase some of the least expensive construction paper, that which will most likely fade in sunlight. If weather permits, place this paper on the ground or on a table in direct sunlight. Have the students place some objects such as saucers or blocks that will block the sunlight from reaching the paper. These can even be done in an artistic manner (to include art in the science lesson). After the Sun has had a few hours to fade the paper, the images will be there to see. This can lead to an investigation with many questions, including:

- Will the Sun create the images if the activity is done inside on a sunny windowsill through glass?
- Will afternoon sunlight work more quickly than morning sunlight?
- Will afternoon sunlight make darker images than early morning sunlight?
- Will sunshine in different times of the year cause different results?
- How do the differences in images change as the length of your shadow changes?

Making "thermosicles" (thermometers immersed in ice cubes) can also be interesting for students to observe. These are made by putting thermometers in an ice cube tray so that the bulbs are in the water, and the register can act as a handle. This will show the temperature of the ice, and it will remain at 0°C even as the ice begins to melt. Thus the child can see the temperature and feel the ice cube at the same time.

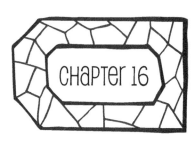
USING THE STORY WITH GRADES 5–8

Children of this age group may well be more adept at using the internet or public press to gather information about the topic addressed in this story. The internet is useful because it is very unlikely that a student will volunteer to take temperature readings at each hour during the day and night (and unreasonable to ask!). Students can keep a record of a few weeks of hour-by-hour temperature cycles in their science notebooks so that they can draw some conclusions from the patterns they see over time.

You may find that giving the probe, "Camping Trip" in *Uncovering Student Ideas in Science, Volume 4*, along with the story as a discussion starter will help your students become more fully immersed in the problem (Keeley and Tugel 2009). The probe may also be used as a formative assessment tool or even as a follow-up tool after the topic has ended.

Children who live above the latitude of the tropics will be familiar with seasonal changes in outdoor temperatures but will still be surprised that the lowest temperatures occur just about dawn. An activity with flashlight and paper can be helpful here. The student holds a flashlight directly over a piece of paper in a darkened room and circles the area that is lit by the beam, which represents the Sun's rays at midday or early afternoon. The flashlight then is shown upon the paper at an angle, (representing dawn) and the lightened area is again circled with a pencil or pen. The area of the light shown at an angle is larger than the area produced by the overhead position of the flashlight. This can be explained to represent the thermal energy received from the Sun as being spread over a larger area and therefore less potent at any given spot within the lighted portion. Heat lamps can also be used with thermometers in the same way to show that temperatures will rise less when the lamp is held at a lower angle than when shone directly down on the thermometer.

When the data are analyzed and the temperatures at dawn are seen to be the lowest, on average, the story can be finished. Dad probably did have someone "knocking on his tent" early in the morning by youngsters who were experiencing the coolest temperatures of the day.

Should the topic of climate come up, information given above might be of value to you. It is not necessarily within the realm of this particular story but the ramifications of climate and climate change are often of interest to students.

RELATED BOOKS AND NSTA JOURNAL ARTICLES

Keeley, P. 2005. *Science curriculum topic study: Bridging the gap between standards and practice.* Thousand Oaks, CA: Corwin Press.

Keeley, P., F. Eberle, and C. Dorsey. 2008. *Uncovering student ideas in science: Another 25 formative assessment probes, volume 3.* Arlington, VA: NSTA Press.

Keeley, P., F. Eberle, and J. Tugel. 2007. *Uncovering student ideas in science: 25 more formative assessment probes, volume 2.* Arlington, VA: NSTA Press.

Keeley, P., and J. Tugel. 2009. *Uncovering student ideas in science: 25 new formative assessment probes, volume 4.* Arlington, VA: NSTA Press.

Konicek-Moran, R. 2008. *Everyday science mysteries: Stories for inquiry-based science teaching.* Arlington, VA: NSTA Press.

Konicek-Moran, R. 2009. *More everyday science mysteries: Stories for inquiry-based science teaching.* Arlington, VA: NSTA Press.

Konicek-Moran, R. 2010. *Even more everyday science mysteries: Stories for inquiry-based science teaching.* Arlington, VA: NSTA Press.

references

American Association for the Advancement of Science (AAAS).1993. *Benchmarks for science literacy.* New York: Oxford University Press.

Childs, G. 2007. A solar energy cycle. *Science and Children* 44 (7): 26–29.

Diamonte, K. 2005. Science shorts: Heating up, cooling down. *Science and Children* 42 (9): 47–48.

Driver, R., A. Squires, P. Rushworth, and V. Wood-Robinson. 1994. *Making sense of secondary science: Research into children's ideas.* London and New York: Routledge Falmer.

Gilbert, S. W., and S. W. Ireton. 2003. *Understanding models in earth and space science.* Arlington, VA: NSTA Press.

Keeley, P., and J. Tugel. 2009. *Uncovering student ideas in science: 25 new formative assessment probes, volume 4.* Arlington, VA: NSTA Press.

Konicek-Moran, R. 2008. *Everyday science mysteries.* Arlington, VA: NSTA Press.

Konicek-Moran, R. 2009. *More everyday science mysteries.* Arlington, VA: NSTA Press.

National Research Council (NRC). 1996. *National science education standards.* Washington, DC: National Academies Press.

Oates-Brockenstedt, C., and M. Oates. 2008. *Earth science success: 50 lesson plans for grades 6–9.* Arlington, VA: NSTA Press.

CHAPTER 17
FROSTY MORNING

ndy turned his 8-year-old body over in his bed as the first rays of daylight peeked into his bedroom window. It looked like another cold morning. He could tell because the birds huddled around the feeder were all fluffed up and looked like they were wearing down vests under their feathers. The bird feeder was hanging from a tree just outside his window and the blowing wind made the feeder sway as though the birds were using it as a swing.

"Cold out there," thought Andy, "and the wind will make it feel colder! Wind does that."

Katie, Andy's older sister who was nine, awoke on the other side of the house in her bed next to a window, which in winter framed the rising Sun. The golden windowpane seemed to brighten her spirits and the warm glow of the Sun made the room feel cozy. "Beautiful day," Katie said to herself. "We'll be able to have outside gym today."

At breakfast, Katie and Andy sat at the table eating their cereal and saying very little. Both children started slowly in the morning and needed a bit of time to get their minds and bodies moving. In fact, everyone was slow this morning. Mom had overslept, the children had dawdled, and the school bus had honked and gone before they could get out of the front door and down the driveway.

"Mom," whined Andy, "we missed the bus. Now what? Do we stay home today?"

"No way!" shouted Katie. "I'm in an assembly play today. I can't miss school."

"I guess I'll have to drive you," said Mother sadly. She had hoped to get in an extra hour's sleep after the children had left since she had been up late the night before. "But, you'll have to help. While I get dressed you two run out and scrape the frost off the car windows, front and back. Katie, you take the back windows and Andy the front. And do a good job!"

The children got the scrapers out of the car trunk and went about their work. The car was parked in the front driveway but it was pulled into a little "L" shaped carport so that the front end was surrounded on two sides by the house and the woodshed. The rear of the car stuck right out into the drive. Andy squeezed through the space between the house and the front fender and prepared to scrape. He could hear the sounds of Katie's scraper as it sliced through the white icing on the rear window.

"Wait a minute," he thought. The windshield was completely clear. "This is cool," he thought with a smile. "Katie is still scraping and my glass is clear as can be. Maybe I won't tell her." But he couldn't resist grinning when Katie came forward and saw that the windshield and front side windows were clear and dry.

She figured it out in a flash. "No fair!' she shouted. "You didn't have to do anything."

"That's right," laughed Andy. "Hope you didn't wear yourself out back there. Mom's going to love the way my windows are clear and shiny. Yours have a lot of streaks on them. Better go back and do a good job."

Just at that moment when Katie was going to demand that Andy go back and finish up on the rear windows, Mom came out of the house and began shooing them into car. "Let's go or you'll be really late."

"Nice job on the windows kids," she said. "The windshield is especially clear Andy. How did you do that?"

"It was easy," said Andy feeling Kate's angry eyes boring into the back of his head.

"He didn't do anything," cried Katie. "His windows didn't have any frost on them at all."

"How could that be?" asked Mom. "One end of the car was frosted and the other one wasn't?"

"Must have been the Sun," said Andy smugly.

"The car was parked completely in the shadows," Katie offered.

National Science Teachers Association

"Well then, the front end of the car was warmer than the back end," replied Andy.

"Why?" asked Mom. "I always thought that the air around the house was the same no matter where is was located. Didn't you?"

"The front end of the car was really close to the house," said Katie. "Could that have anything to do with it?"

"Would that make the temperature there different than any other place around the yard?" asked Andy. "Are there places in our yard or around our house that have different temperatures?"

"I guess there are ways to find out," said Mother. "But right now, we need to get you guys to school."

PURPOSE

The theme of the story can be summed up in one word: microclimates. Have you noticed that there are variations among the temperatures broadcasted on your radio or TV and your own thermometer? Have you noticed that there are differences in temperature depending upon where you place your thermometer? Have you noticed that certain plants do better or worse depending upon where in your yard they are planted? This story highlights the fact that there are physical differences in every location that modify the amount of heat that is distributed in and around that location. This creates little islands of climatic diversity called microclimates. This story should result in deeper understanding of the effects of differential heating of the Earth's surface by the Sun.

RELATED CONCEPTS

- Atmosphere
- Solar energy
- Heat
- Melting
- Patterns
- Mass
- Radiation
- Absorption
- Temperature
- Climate
- Differential heating

DON'T BE SURPRISED

There are several questions in the "purpose" paragraph above. If you were to take a thermometer outdoors with you and take temperatures at different places around your home or school, you might be surprised at the differences you would find. Your students will probably predict double-digit differences but will soon be satisfied with a few degrees. They will be surprised at the higher temperatures near the foundation of homes or on or above patches of bare soil. They might predict that areas protected from wind would also be warmer than open spots but not that temperature readings high above the ground may be different than those close to the ground.

CONTENT BACKGROUND

Climate can be defined as the prevailing weather conditions in an area over a long period of time. People who live in desert areas expect to have warm days, cool nights, low humidity, and few changes in weather patterns over the year. Those in coastal areas expect high humidity and winds that change from offshore to onshore as the day progresses. In extreme southern and northern areas of the globe it is normal to expect cold winters and warm summers. People who live on

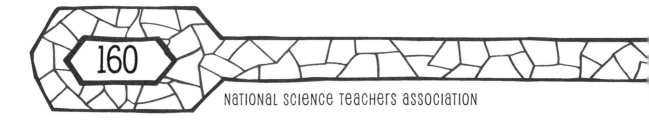

islands can expect that the surrounding water will moderate the temperature and humidity over the year. Of course there are always exceptions such as storms that bring extremes to any type of climate, but overall, the climatic conditions change very little from decade to decade. On the Earth, there have been dramatic climatic changes over millions of years due to shifting continents and cataclysmic conditions such as huge volcanic eruptions and meteors striking the planet. But, over scores of generations climates remain reasonably steady and predictable.

Within climates however, there are microclimates, which may affect entire cities and counties. San Francisco, for example, due to its proximity to the ocean and warm currents, its hilly landscape, and prevailing winds, has whole sections of the city with different microclimates. San Diego has a similar situation. Cities are a dynamic example of human-constructed microclimates. The use of concrete in buildings and streets provides a heat-absorbing mass that collects heat and radiates it into the atmosphere. Thus, cities are generally warmer than an open countryside. The Redwoods National Park is located where it is because of the microclimate that provides a daily fog on the shoreline, which in turn provides the necessary moisture for the huge redwood trees, which grow only there. The microclimate around your school or home is not necessarily as dramatic but does provide a surprise to most people who take the time to map the differences in climate in their little piece of the world.

My wife and I added about 50 centimeters of depth in black soil to our raised beds in our New England vegetable garden during the fall months and then added a plastic cover over the bed. We were harvesting lettuce in December even after several hard frosts. The added mass of soil, combined with the trapping of solar energy, warmed the soil the several degrees necessary to keep the environment warm during the cold nights and created a little greenhouse effect and a small microclimate.

You will certainly find areas where there is more sunlight during the day than in other areas. There will be areas where there is more moisture than others. All of these factors will change the climate in those places and provide you with help as well as challenges when it comes to choosing the best place for certain plants. We tend to think of our land as homogeneous when it comes to temperature, humidity, and climate but this is not the case. It may appear from illustrations of the Sun shining on the Earth that all areas receive equal heating but the planet is made up of many different kinds of materials and landforms. The same is true of our smaller, more familiar areas. Thus, Andy and Katie found that when the car was parked in the little protected area, the microclimate surrounding the car provided enough heat from the house side to melt the windshield but not the rear window a mere three or four meters away.

We also create microclimates in our homes by insulating walls and roofs and adding humidifiers to increase the relative humidity in the home. Thermostats are placed in strategic places so that they are most likely to record the temperature and regulate the heating system in a comfortable way. One does not, for example, place a thermostat directly over a radiator because that does not give an accurate reading of the general temperature in the house. They are not placed on the ceiling or on the floor for the same reason. They usually are placed at eye level, which is where we generally spend our time.

During the last half of the 20th century and the first part of the 21st century, there has been a great deal of concern about global warming. Due to our industries and automobiles it is feared that "greenhouse gases," mostly carbon dioxide, have put a layer of gas into our atmosphere much like the plastic cover in our garden vegetable bed that is causing the average atmospheric temperature of our earth to rise slowly but steadily. Ice caps are melting and there is fear that the additional water in the oceans will raise the sea level and flood shoreline communities and endanger species of animals such as polar bears in the Arctic. This theory is garnering support by scientists all over the world, and if true, global warming will cause megaclimatic changes everywhere.

Most of our weather takes place in the layer of air (the troposphere) that ranges from sea level to about 13,000 meters although some storm clouds can project above that level. But in this story we are concerned about the layer of air just a meter or two above the surface.

related Ideas From The National Science education standards (Nrc 1996)

K–4: Changes in Earth and Sky
- Weather changes from day to day and over the seasons. Weather can be described by measurable quantities, such as temperature, wind direction and speed, and precipitation.

K–4: Objects in the Sky
- The sun provides the light and heat necessary to maintain the temperature of the earth.

5–8: Structure of the Earth System
- The atmosphere is a mixture of nitrogen, oxygen, and trace gases that include water vapor. The atmosphere has different properties at different elevations.

5–8: Earth in the Solar System
- The sun is the major source of energy for phenomena on the earth's surface, such as growth of plants, winds, ocean currents, and the water cycle.

related ideas from Benchmarks for science literacy (aaas 1993)

K–2: Energy Transformations
• The sun warms the land, air, and water.

3–5: Energy Transformations
• When warmer things are put with cooler ones, the warm ones lose heat and the cool ones gain it until they are all at the same temperature. A warmer object can warm a cooler one by contact or at a distance.

6–8: Energy Transformations
• Heat can be transferred through materials by the collision of atoms or across space by radiation.

USING THE STORY WITH Grades K-4

Grades K–2 students often have a very difficult time reading thermometers. Their most common experience with thermometers is probably having their temperature taken. Using this story with early childhood students would offer an opportunity to learn about the thermometer and to talk about being hot and cold and what conditions contribute to that feeling. If they are able to use the thermometers, they can certainly keep records of the indoor and outdoor temperatures from day to day and use these records to study the variations in readings. The question arises about the use of Celsius scales for young children. It seems that if children can become bilingual, they can also become comfortable with two scales of temperature. You must make sure that they realize that the difference in numbers is merely due to the kind of thermometer being used rather than the temperature they are measuring. You should not spend time teaching children how to convert from one scale to another but encourage them to "think" in one scale or the other.

Soon they will see that energy from the Sun causes higher temperatures in sunny places. Likewise, shady places receive less energy from the Sun, so they have lower temperatures. They can be introduced to the idea that the Sun is the source of warmth for the Earth and can verify this with their newly found skills in temperature reading. Keeping a daily record of the weather is a valuable skill, especially if it includes connecting their data to other forms of weather observation such as clouds, precipitation, and overall weather conditions. This story has been specifically tested with third graders and older children with good results.

USING THE STORY WITH GRADES 5-8

As with any story, the lesson begins with a discussion of the mystery. The students will probably have personal stories that will parallel the experiences of the story characters. Asking the students to tell you what they believe about how temperatures can vary in different places will reveal what kind of ideas they are bringing to your classroom. Your question could be something like, "What could cause the temperatures to be so different at the two ends of the car?" Once their ideas are recorded on the class chart and in their science notebooks, the "I think…" statements can be changed to questions for testing. You may also ask them if they think that there are different temperatures in various places around the school grounds, why they think so, and how they could find out. Combining their experiences and your questions should help students develop a series of plans to map the school grounds with thermometers. You may also find this to be an opportunity to combine mapping skills and measurement skills into the unit, thereby integrating your math and science curricula. Once the mapping is done, and students have had a chance to observe the school grounds more carefully, they may be able to develop some hypotheses about where they believe the temperatures might vary. At first, the majority will focus on the sunny spots with obvious results. Remind them by returning to the story that Andy and his sister found that the carport area was not in the Sun, yet had variations in the temperature within a small area. This may encourage them to seek out similar spots on the grounds to see if they get similar results. In their science notebooks, they should be describing the areas they are sampling with notations about the shape of landforms, human-made constructions, or changes in the topography and ground cover. They may note for example, that the morning Sun shines consistently on brick or concrete areas of the building and that even later in the day, those spots are still warm even though they are no longer in direct sunlight. This may lead you to gather data in various locations at different times of the day if this variable comes up.

This may also lead to some additional investigating into the heat-absorbing qualities of various materials such as water, concrete, asphalt, soil, grass, and so on. Further experimentation on how quickly and how long these materials retain heat will probably result in additional questions and findings. Is there a difference between how long it takes various materials to reach a certain temperature in direct sunlight? Do some materials stay warmer than others over time?

It is important to note that the questions and experiments are now in the hands of the students, and their notebooks should reflect this. Remember to ask for statements of support for any conclusions they make. Now, they may be ready to finish the story, and you will be ready to see the amount of growth they have made.

NATIONAL SCIENCE TEACHERS ASSOCIATION

related BOOKS and NSTA JOURNAL articles

Driver, R., A. Squires, P. Rushworth, and V. Wood-Robinson. 1994. *Making sense of secondary science: Research into children's ideas.* London and New York: Routledge Falmer.

Keeley, P. 2005. *Science curriculum topic study: Bridging the gap between standards and practice.* Thousand Oaks, CA: Corwin Press.

Keeley, P., F. Eberle, and L. Farrin. 2005. *Uncovering student ideas in science: 25 formative assessment probes,* volume 1. Arlington, VA: NSTA Press.

Keeley, P., F. Eberle, and J. Tugel. 2007. *Uncovering student ideas in science: 25 more formative assessment probes,* volume 2. Arlington, VA: NSTA Press.

Robertson, W. 2002. *Energy: Stop faking it! Finally understanding science so you can teach it.* Arlington, VA: NSTA Press.

references

American Association for the Advancement of Science (AAAS). 1993. *Benchmarks for science literacy.* New York: Oxford University Press.

National Research Council (NRC). 1996. *National science education standards.* Washington, DC: National Academies Press.

CHAPTER 18
MASTER GARDENER

Eddie's mother was a professional gardener. She had a wonderful reputation in her business, Everything Green. People said she could make plants grow in the middle of the interstate highway. Of course this wasn't true but it showed how much people thought of her work. In fact she was out on a job the day the phone rang and Eddie answered it.

"Hi Eddie, I need to talk to Kerry about an errand," his Mom said.

Eddie handed the phone to Kerry. Sam's brother Kerry was 16 and could drive so he got lots of calls to help his Mom, especially during the planting season.

"Hi Kerry, I need you to do me a favor. I need you to take the truck and go to the farmer's supply store and get me two 80 lb. bags of coarse sand. That's coarse sand, the sand with big grains," she emphasized. "If they don't have coarse sand, go to the garden center and try there! I also need a big bale of peat moss. I'm over at Mrs. Brown's on Amity Street. Can you do that right away?"

"Sure Mom, be there in half an hour," said Kerry. He was a big shot since he could drive the pick-up and do things like that. As he was hanging up, he said to Eddie, "Hey little bro. Want to come along and see how the big folks work?"

Eddie ignored the little bro comment and said sure. He had nothing else to do anyway and thought that he might get Kerry to buy him an ice cream if he played his cards right.

Kerry got the coarse sand at the farmer's supply store and then headed for the garden center for the peat moss.

"Mom must have a real problem garden if she needs this stuff right away," said Kerry.

"What's a problem garden, anyway?" asked Eddie.

"She probably dug down and found lots of clay and needs to rebuild the soil."

"How can you rebuild soil? Soil is soil, isn't it?"

Kerry breathed a long sigh of impatience and told Sam that he should ask his mom when they got there. Eddie did just that.

"When I find a garden with too much clay in the ground, the water won't drain right so I have to rebuild the area by adding things that will break up the clay," his mother explained. "I add coarse sand to allow water to drain and then good rich soil and then peat moss to hold moisture and lighten the soil."

In the meantime, Kerry was lugging the sand and peat moss over to the garden where his mom was working. Mrs. Brown always composted her garbage and so there was plenty of good stuff in the compost bin.

Eddie's mom dug the clay up; broke it up with her shovel; added the sand, compost, and peat moss; and mixed it up really well with her rake.

"There, now that should take this hydrangea and make it happy," she panted.

On the way home, Eddie asked Kerry if he knew where all of this stuff, coarse sand, peat moss, and everything came from. "I thought dirt was dirt," Eddie said.

"Not sure, little bro. Better ask Mom. All I know is that I was told not to bring fine sand and I thought...'uh, sand is sand.' Must be different stuff on this old Earth that we don't know about. I thought sand was just...you know, little rocks. Maybe the different kinds of dirt come from different places. Maybe the sand washes up from the ocean in a different way to get to be coarse."

Eddie was puzzled. There was no ocean here. There are different kinds of sand? All soil is different? You can rebuild soil? If the answer to all of these questions were yes, then where did all of this different stuff come from and where would it finally end up? Isn't all sand alike? Isn't all soil alike? What is soil anyway and where does it come from? Now Eddie was really puzzled!

National Science Teachers Association

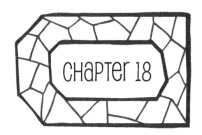

PURPOSE

My grandmother, who was a farmer, often heard people talk about soil and call it dirt. Her response was, "Dirt is stuff where it doesn't belong. Soil is where we grow our crops." This story is designed to spur an inquiry activity about the process of weathering and soil formation. Evidence lies all around us if we look closely enough and ask the right questions. Eddie is helping us by asking some of these questions and we would like the children to ask some of their own questions as well. The result should be a better understanding of the materials that make up our planet and how they came to be.

RELATED CONCEPTS

- Rocks
- Deposition weathering
- Soil
- Minerals
- Erosion
- Decomposition

DON'T BE SURPRISED

Most students and a great many adults have the same questions that Eddie voices. "Isn't all soil alike?" "How can you *rebuild* soil?" The things we see everyday become so familiar that we no longer really "see" them or ask questions about them. I once visited a beach in Cornwall, England, sat down in the "sand" and began to allow it to filter idly through my fingers as I viewed the scenery around me. Imagine my surprise when I finally realized that a lot of this particular "sand" was made up of tiny shells of formerly living sea creatures as well as regular sand. I wondered how many people had visited that beach and never realized that they were basking on tiny seashells. When students actually begin to analyze sand and soil by looking closely at these substances with magnifying glasses or microscopes, their own questions will begin to form. We take soil for granted as we do sand. Sifting through soil samples from various locations will amaze children when they discover the many wonderful organic and inorganic materials that make up this common substance. There will be tiny bits of rocks and minerals, skeletons of once living creatures, minute living creatures, and as one little first grader exclaimed to me, "I found a bird's toenail!"

These observations will give rise to questions about how these little bits of rocks and minerals got to be so small, and that leads into the whole misunderstood area of weathering, erosion, decomposition, soil formation, and the importance of all of these processes in the Earth's past and future. Children and adults alike do not understand decomposition and the processes that return organic material to the soil. They believe that an apple that drops to the ground just disappears or miraculously becomes soil. Perhaps probing into the questions raised by this story will help raise their awareness of this common treasure.

CONTENT BACKGROUND

Rocks are made up of minerals. And what are minerals? Simply stated, mineral is a name given to substances that are not animal or vegetable. (Remember the question asked in the game 20 Questions "Is it animal, vegetable, or mineral?") Scientists believe that over the billions of years since the planet was formed, a process we now call the rock cycle has taken place countless numbers of times. Rock was originally formed by the volcanic actions so prevalent on the newborn planet that it was made entirely of molten lava. This lava contained all of the dissolved minerals we have in today's world. Over eons of time the Earth cooled, leaving us with a barren planet covered with this cooled lava, which formed igneous rock. This is going on today in Hawaii, or the Big Island as it is called. Volcanoes originally formed the Hawaiian Islands from deep below the ocean surface and these volcanoes, now above sea level, continue to spew out lava building more land and the Big Island grows larger each day. In fact, deep within the sea to the southeast there is another island forming from an undersea volcano, which some centuries from now will become another new Hawaiian island when it reaches the ocean's surface.

When the Earth had cooled sufficiently to allow water to exist in liquid form, the seas and lakes formed and the water cycle (see the story in chapter 14, "The Little Tent That Cried") kept Earth's water cycling from land to sky and back down again as rain. The volcanically formed mountains were broken down by weathering and water was able to run downhill to the seas, lakes, and rivers, carrying with it the broken pieces of the mountain in a process called erosion. Let's take a look at weathering and erosion and point out the differences between the two processes since they are often thought to be the same.

Weathering is a term that refers to the breakdown and changing of rocks and minerals. The processes causing weathering take place at the Earth's surface and are broken into three main types: chemical, physical, and biological.

When rocks undergo a chemical change, there are changes in many of the minerals in the rock. Acid in rainwater may come in contact with the minerals, or rainwater may react with the minerals to form an acid. Some of the compounds and atoms in the mineral may actually go into solution in the water and be removed completely or the addition of oxygen may cause the mineral to flake off more easily. The higher the climatic temperature and the more moisture there is, the quicker the processes continue to work.

When mechanical weathering takes place, the rock is broken into pieces by actions such as scraping or grinding, repeated freezing and thawing resulting in expansion of water in cracks in the rock, and differential heating and cooling. All of these processes may cause wearing down of the surfaces or the flaking off of surfaces from the larger rock. Glaciers do their bit of grinding as well, thousands of years ago as well as today in places where glaciers still exist. Usually no chemical changes take place.

Biological weathering disintegrates rock and mineral particles due to the action of biological organisms. Examples include simple breaking down of rock particles by passing through an earthworm's digestive tract. Burrowing animals can exert pressure

170

on rocks and bring them to the surface where they can be acted upon by other forces. Bacteria and fungi can cause chemical weathering by the chemicals they produce. Pressure exerted by growing roots can produce enormous forces. Perhaps you and your students have seen cement sidewalks pushed up or broken by tree roots. Plants can grow in cracks of rocks and cause the rocks to break up as the plant grows. And finally the carbon dioxide given off in cellular respiration mixed with water can lead to a chemical weathering by acid wearing away the surface of rocks and minerals.

All of these processes tend to weaken or break off parts of the larger rocks so that they can become prey to the process of erosion. As the rock particle is loosened or broken loose from the parent rock, it can move. It usually moves downward due to gravity. Erosion is defined as the movement of rock or soil particles down a slope moved by some substance such as water, air, or ice and it may cause ruts or channels. If a particle is loosened from a rock and stays where it is, it is called weathering. Once the particle begins to move due to a flowing agent, it is erosion. The Grand Canyon was caused by both weathering and erosion. Chemical, physical, and biological processes of weathering loosened the particles of rocks so that the Colorado River could carry the particles off by erosion, cutting a canyon a mile deep into the surrounding rock.

These particles of the original rocks found their way into the seas and were eventually washed up on the distant beaches of the ancient seas. Most of the sand on the world's beaches is made up of quartz, the most common mineral on the face of the Earth. These formed layers of sediment (as sand), eventually were buried, baked, and turned into sedimentary rock by combination of the heat and pressure of the land above them. This sandstone may have been weathered away again or it may have been buried and heated and squeezed further to form metamorphic rock. When life formed in the seas and the skeletons of the animals and plants formed layers of sediment, they became limestone and eventually marble if pressure was applied before it became exposed. As the earth continued its unrest and rose and fell, forming more mountains and in some cases lifting the newly formed rock to mountaintops, the process began all over again. And on it goes to this day but we see only the results, because the process is exceedingly slow and cannot be seen even in many lifetimes. The process that produced the rocks, minerals, and soil we see today has been responsible for recycling the original material from Earth's origin and continues the cycle into the future.

However, we can see one process that is ongoing, namely the formation of soil and its continual enrichment. As the small particles of rocks and minerals are broken down further, they are mixed with dead plant and animal material, broken down to their basic components by the action of decomposers like bacteria, fungi, lichens, and larger animals like earthworms and insects that live in what is now called soil. So in a metaphorical way, soil is a living organism, forever changing and recharging its nutrients so that plants and fungi can grow and gain mineral nourishment and replenish that which is used by growing, living things. Each time a once-living thing enters the soil it is broken down into its most basic components by decomposers and keeps the soil rich and fertile. Eddie's mother refers to the fact that her client composts and therefore has good soil that she can use. Her client composts by putting only the vegetable matter from her garbage into a container along with water, soil, and the organisms that break down the vegetable matter. The person who composts is feeding the organisms in the soil so they can break down the vegetable material into nutrients that can be used by plants. Remember, these nutrients are not food but include elements such as copper, magnesium, potassium, and phosphorus, as well as nitrogen and calcium that are necessary for plants to function. Gardeners often add fertilizers containing these necessary nutrients to the soil. These are erroneously called "plant food," but are more like the vitamins and minerals we take to supplement our own food intake. Since plants are producers, they make their own food, so speaking of fertilizers as plant food is incorrect and confusing.

Soil can be one of three main types: sandy, silty, or clay. Besides the living things in the soil there are rock and mineral particles that differ in size and structure. Sandy soil is composed of larger particles with lots of space between them, which allow great drainage but retain little water and dissolved minerals. Silty soil is made up of much smaller particles visible only by microscope and is formed by mechanical weathering. It is often seen as dust and can be blown by wind for miles.

Clay is formed mainly by chemical weathering and is made up of tiny closely packed particles visible only through an electron microscope. It has poor drainage and is structured in layers that make it difficult for plants to penetrate.

Soil has texture and structure. The texture of the soil is determined by the proportions of sand, silt, and clay and by which one dominates the properties of the soil. You cannot easily change the texture of the soil, but you can change the structure of the soil, which is the arrangement of the different types of soil it contains. Eddie's mom changed the structure of the soil, dominated by the clay texture, by adding organic material such as peat moss to hold water, compost to add nutrients, and sand to improve drainage. Good growing soil is loose and crumbly and takes in water easily, allows air to move in and out, and allows for plant roots to penetrate easily finding the water and nutrients they need. Good growing soil does not form thick, heavy clods. Eddie's mom did just this as she rebuilt the soil structure.

related ideas From the National Science Education Standards (NRC 1996)

K–4: *Changes in Earth and Sky*
- The surface of the Earth changes. Some changes are due to slow processes, such as weathering and erosion.

K–4: *Properties of Earth Materials*
- Earth materials are solid rocks and soils, water and gases of the atmosphere. The varied materials have different physical and chemical properties, which

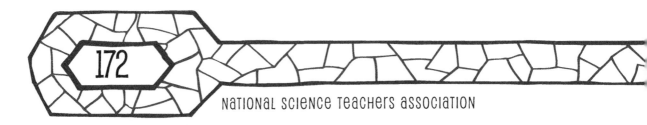

make them useful in different ways, for example, as building materials, as sources of fuel, or for growing the plants we use as food. Earth materials provide many of the resources that humans use.

- Soils have properties of color and texture, capacity to retain water and ability to support the growth of many kinds of plants, including those in our food supply.

5–8: *Structure of the Earth System*

- Soil consists of weathered rocks and decomposed organic material from dead plants and animals and bacteria. Soils are often found in layers with each having a different chemical composition and texture.
- Landforms are the result of a combination of constructive and destructive forces. Constructive forces include crystal deformation, volcanic eruption, and deposition of sediment, while destructive forces include weathering and erosion.
- Some changes in the solid earth can be described as the "rock cycle." Old rocks at the Earth's surface weather, forming sediments that are buried, then compacted, heated, and often recrystallized into rock.

related ideas from Benchmarks for science literacy (aaas 1993)

K–2: *Processes That Shape the Earth*

- Chunks of rock come in many sizes and shapes, from boulders to grains of sand and even smaller.

3–5: *Processes That Shape the Earth*

- Waves, wind, water, and ice shape and reshape the Earth's land surface by eroding rock and soil in some areas and depositing them in other areas, sometimes in seasonal layers.
- Rock is composed of different combinations of minerals. Smaller rocks come from the breakage and weathering of bedrock and larger rocks. Soil is made partly from weathered rock, partly from plant remains—and also contains many living organisms.

6–8: *Processes That Shape the Earth*

- Although weathered rock is the basic component of soil, the composition and texture of soil and its fertility and resistance to erosion are greatly influenced by plant roots and debris, bacteria, fungi, worms, insects, rodents, and other organisms.

- Some changes in the Earth's surface are abrupt (such as earthquakes and volcanic eruptions) while other changes happen very slowly (such as uplift and wearing down of mountains). The Earth's surface is shaped in part by the motion of water and wind over very long times, which acts to level mountain ranges.

USING THE STORY WITH GRADES K-4

I would like to suggest that you consider using the probe, "Beach Sand," in the book *Uncovering Student Ideas in Science, Volume 1* (Keeley, Eberle, and Farrin 2005) even before reading this chapter's story to your students. You will learn what kinds of preconceptions your students are bringing to your class and which ones will have to be addressed.

One of the biggest problems in studying this topic with young children is the time scale involved. How many of us can really comprehend a million of anything, much less a billion? The wearing down of mountains must seem like an impossible fairy tale to a young child, but they can observe the particles of rocks and minerals in common objects like soil and sand. After reading the story, children are often anxious to take a look at different types of sand and soil. If you have different gauges of strainers, the children can separate the different sizes of sand by sifting them. But one of the most effective ways of beginning with young children is to let them use magnifiers and toothpicks to work their way through a small amount of soil taken from the garden or the school property. They may need to have help learning how to use the magnifiers or you may use the type held on tripod stands. Using a magnifier involves holding the glass up to the eye and bringing the object to be observed up to the eye rather than putting the magnifier down near the object. My experience is that children will enjoy finding things in the soil and trying to identify them. Of course these should be recorded in their science notebooks with plenty of labeled drawings. The analysis of soil should reveal that it is made up of many different things, including living and once-living material, and nonliving material such as water and tiny particles of rocks and minerals.

Using the same techniques with a small helping of sand is also rewarding and asking the children to sort, with the aid of a magnifier, the different types of grains in piles that "go together" will show them that sand is a mixture of many beautiful and newly discovered little rocks and minerals of amazing varieties of size, shape, color, and luster. If possible, try to obtain sand samples from several different locations. Even though quartz will usually be the dominant mineral, sand from different locations can be sharp or rounded or fragmented. If you get coral sand or green or black sand, the differences will be dramatic. There may even be a few little seashells. Comparing these grains to larger rocks that contain the same kinds of minerals and crystals will usually convince students of the origin of the sand. During discussions, students are made aware of the continuity between the small grains and the larger rocks and boulders they may see in the naturally visible world. A 10-minute field trip to find rocks and

174

minerals in the natural world is well worthwhile and broadens their understanding of the connections to larger landforms. And in this connection you may help them to see that the wearing down of boulders into sand will take a long, long time. For young children this is a first step toward understanding the time scale involved.

Third and fourth graders can also benefit from some of the same kinds of activities as described above. One can expect a bit more sophistication in their recording, and drawings of what they find. They can also learn to categorize these rocks by color, luster, and hardness, and some students may benefit from learning the differences between igneous, sedimentary, and metamorphic rocks. There are many commercial units on rocks and minerals for follow-up work, but the inquiry they do on sand and soil will prepare them better for working with these programs of identification and categorization of rocks and minerals in later grades.

I might suggest that you consider planting seeds in different kinds of soils, keeping all other variables constant to see if there are differences in germination ratios, germination time, health of plants, and size of plants, and so on. Older students might also use kits that determine the amount of various minerals in types of soils that affect plant growth.

USING THE STORY WITH GRADES 5–8

I would again suggest that you consider giving your students the probe "Beach Sand," found in *Uncovering Student Ideas in Science, Volume 1* (Keeley, Eberle, and Farrin 2005). You may also want to use the probe "Mountain Age," in the same book. These probes will provide you with knowledge of the preconceptions your students bring to class on the topic of the rock cycle.

Upper elementary-level and middle school students may have already done some study of weathering in earlier grades. The probe will give you an idea of how many of their preconceptions were changed by past instruction. Their conception of long periods of time will probably be more mature, but they will need to explore the questions Eddie has about the origin of rocks, sand, and soil by developing questions of their own. Mark Girod, in his article "Sublime Science," in the February 2007 issue of *Science and Children* gives some wonderful suggestions on how to help children understand the magnitude of large numbers. By using pages printed with 10,000 dots (done on the computer), he was able to show his students large numbers, and through a series of activities involving counting and the guessing of numbers the students were able to actually see what large numbers of objects looked like. Imagine, 10 sheets of 10,000 dots actually showed students what a million looked like.

After reading the story, you may want to develop a chart of their "best thinking" for periodic review as they perform their inquiry. The gardening theme of the story is aimed at making them wonder about the various types of materials on the Earth. If you are fortunate, you may already have a "rock hound" or two in your class. If this is the case, you may have direct access to a lapidary rock tumbler. This small device, which

is usually used to wear down and polish rocks and minerals into beautiful stones, can also be used to show in quick time the process of weathering that takes centuries in the real world. Your local Earth science teacher may have a tumbler you can borrow. They are quite inexpensive and can be usually found in local craft stores.

Your students' inquiry questions may have more to do with comparing different kinds of sand from various locations or comparing different types of soil or even doing a survey of biomass (available natural food, like insects or plant matter) in various soils, than they do on the age of the Earth.

One must always be aware of the possibility of children coming from homes where biblical stories are considered to be literal truth. These children will believe that the Earth is only 6,000 years old and base their beliefs on faith alone. While we, as teachers, cannot argue science versus faith questions, we can state specifically that the numbers science considers to be true are based upon scientific tests and verified theories. There is no use arguing science and religion in a science classroom since they are based on two entirely different systems of belief and verification. This may also give you an opportunity to discuss the nature of science with your students if this seems appropriate.

You may also consider using a stream table, which allows the students to simulate, in quick time, the erosion and deposition of soil or sand by running water. Plans for simple stream tables can be found easily on the internet. Care must be taken to emphasize that the erosion sometimes seen in a stream table may take much longer to accomplish in the real world. They should be aware that natural disasters like tsunamis, hurricanes, and tornadoes might change the landscape very quickly but that the normal cutting of canyons and valleys and the creation of such landforms as the Devils Tower take thousands if not millions of years. The stream table is only a model used to test out theories to see if they can prove to be true in a longer time frame.

Thinking and learning about the enormous time frame for the building and rebuilding of Earth's landforms may seem to be a daunting task, but it is a particularly important concept for students to master, especially since we are capable of modifying so many parts of Earth through our negligence and hurry to change the planet for our own benefit and greed. One can never go wrong in teaching our children about their responsibility as stewards of our natural resources.

related BOOKS and NSTa Journal articles

Coffey, P., and S. Mattox. 2006. Take a tumble. *Science and Children* 43 (7): 33–37.

Driver, R., A. Squires, P. Rushworth, and V. Wood-Robinson. 1994. *Making sense of secondary science: Research into children's ideas.* London and New York: Routledge Falmer.

Gibb, L. 2000. Second-grade soil scientists. *Science and Children* 38 (3): 24–28.

Girod, M. 2007. Sublime science. *Science and Children* 44 (6): 26–29.

Keeley, P. 2005. *Science curriculum topic study: Bridging the gap between standards and practice.* Thousand Oaks, CA: Corwin Press.

Keeley, P., F. Eberle, and L. Farrin. 2005. *Uncovering student ideas in science: 25 formative assessment probes,* volume 1. Arlington, VA: NSTA Press.

Keeley, P., F. Eberle, and J. Tugel. 2007. *Uncovering student ideas in science: 25 more formative assessment probes,* volume 2. Arlington, VA: NSTA Press.

Laroder, A., D. Tippins, V. Handa, and L. Morano. 2007. Rock showdown. *Science Scope* 30 (7): 32–37.

Levine, I. 2000. The crosswicks rock caper. *Science and Children* 37 (4): 26–29.

McDuffy, T. 2003. Sand, up close and amazing. *Science Scope* 27 (1): 31–35.

Sexton, U. 1997. Science learning in the sand. *Science and Children* 34 (4): 28–31; 40–42.

Verilar, M., and T. B. Benhart. 2004. Welcome to rock day. *Science and Children* 41 (4): 40–45.

references

Gibb, L. 2000. Second-grade soil scientists. *Science and Children* 38 (3) 24–28.

Keeley, P., F. Eberle, and L. Farrin. 2005. *Uncovering student ideas in science: 25 formative assessment probes,* volume 1. Arlington, VA: NSTA Press.

A DAY ON BARE MOUNTAIN

Bare mountain is a mere hill compared to the Rocky Mountains or the Sierras or even the relatively small White Mountains of New Hampshire. The top stands only 1,000 feet above the valley below. But on a clear day, you can get a beautiful view of 30 miles or more from the topmost overlook. It is named Bare rather than Bear because of its bald, treeless top.

One beautiful, crisp, autumn day, Mari and Kerry decided to take a hike up the mountain to view the New

England display of fall leaf colors from the top. They packed some sandwiches so they could have a picnic on the summit, put a leash on their dog, Ticket, and headed out for the trailhead.

At the bottom of the mountain, they saw the familiar patch of fine sand that they jokingly called "the beach." Mari's low shoes soon became loaded with sand and she complained to Kerry. "I wonder where this sand came from? It's the only place on this trail where I have to empty out my shoes before I can go on!"

"Yeah, I know," said Kerry. "You would think that there should be a lake or something here to go with the beach sand."

In a few minutes, they were walking among huge stair step–size boulders as they climbed up the steep trail.

"Now this is more like it," said Kerry as she took giant steps over the smooth, rounded boulders that took her higher and higher. "I'll bet these boulders probably came from up above and fell down here."

"And boulders don't get in your shoes!" laughed Mari.

About halfway up the mountain they came to a place they had encountered often on this climb. The trail went straight ahead and then looped around to the right. On the right-hand slope, there was a patch of smaller, loose, flat rocks that led directly to the trail up above and where they would be walking in about 10 minutes.

"Looks like we could take a shortcut and go up that slope. We'd save about five minutes, and it's not too steep. Let's try it," said Kerry.

"I don't know," said Mari. "It looks a little slippery and dangerous."

"Aw, come on, sis," urged Kerry. "It looks like fun."

But it wasn't! For every step they took forward on the small flat stones, they seemed to slip back at least a half step. They couldn't seem to get a grip with their feet and the 5 minutes they wanted to save turned into 20 minutes of slipping and sliding until they reached the bushes and trees at the edge of the trail. They had to pull themselves up by the tree branches to arrive finally on the trail, panting and sweaty.

Ticket was waiting patiently for them at the top. She hadn't had any trouble, but then she had "four paw drive," as Kerry put it.

"Okay, Mari, you were right. That was a tough climb. I guess the sand at the bottom of the trail wasn't as much trouble as these little rocks. I wonder where they came from?"

"Well, if the sand at the bottom came from the top, these probably did too. But how come they are so much bigger and landed here on this slope?"

"Yeah, and there are some more on the trail below," observed Kerry, "but they weren't a problem on the flat part of the trail."

They decided to stay on the trail from then on. Soon they were walking past huge boulders and then, at last, up the final 100 meters to the top. This part of the trail also had a lot of the loose stones, but they had been pounded into the soil by thousands of hikers and seemed a lot smaller. The sisters reached the top and walked over to view the valley on the peak's solid rock surface that was half the size

of a football field. There were a lot of cracks filled with water, and Ticket lapped thirstily at her natural drinking bowls in the rock as Kerry and Mari drank from their bottles.

As they looked out over the valley, they wondered about all of the rocks of different sizes down below. Why was the sand only at the bottom and where did those smaller slippery stones and then those big boulders come from? Why were there steep cliffs on all sides of this mountain? Was this an old mountain that was falling apart? And how come the top had all of those cracks yet was in one huge piece? And the biggest question of all, where did this mountain come from in the first place when all around it is nothing but valley almost as far as they could see?

PURPOSE

This story brings up questions about the geology of mountains and the weathering and erosion that takes place as nature breaks down the higher landscape until it is eventually level. It also leads to inquiry into the origin of mountains.

RELATED CONCEPTS

- Weathering
- Mountain building
- Erosion
- Rock cycle

DON'T BE SURPRISED

Students may have the idea that all mountains are volcanoes and were formed by eruptions. Some believe that mountains are clumps of dirt that are just higher than the surrounding landscape. Some students think (perhaps through the influence of religious creeds) that the Earth was formed just a few thousand years ago and that it has always looked like it does now. For most young children, the idea of change in the landscape that has occurred over millions of years is totally beyond their comprehension, as it is also for many adults.

CONTENT BACKGROUND

Geology, or the study of the structure of the Earth, is an important area of investigation for students because it examines a part of our world that is all around us and very vital to our everyday existence. Geology ties together many other sciences including biology, physics, and chemistry. Geology is based on the idea that changes occur over long periods of time. The Earth is thought to be approximately 4.5 billion years old. This estimate is based on scientific techniques involving radioactive decay. Those known as the "Young Earth advocates" dispute this on the basis of scriptural and questionable scientific critiques, and insist that the Earth is no more than 6,000 years old. My intent here does not include getting involved in a creationist-evolutionist debate, but I hope to offer geological principles and hypotheses that are supported by scientific evidence.

Bare Mountain really does exist in western Massachusetts and is hiked daily by dozens of people. It was formed during the time when Pangaea, the supercontinent (more about this later), broke apart and the plates moved to their present positions. Originally, Bare Mountain was formed of igneous or molten rock. Later, forces within the Earth transformed the mountain, which makes it an example of *fault blocking*. When large fractures in the Earth's surface move upward or downward due to forces within the Earth, sharp mountain structures are created. They usually have deep, steep sides when they are first formed. These, as well as the

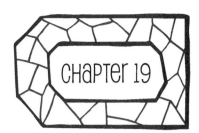

sharpness of their peaks, are moderated by weathering and erosion over time—meaning millions of years.

The differences between *weathering* and *erosion* are distinct. Weathering is the breaking up of rocks and minerals by biological, chemical, physical, and human action. When these broken pieces are carried away by wind or water, this is called erosion. Obviously, weathering occurs first, followed by the erosion of the surface where the weathering has taken place. An epic example of this is in the Grand Canyon, where the weathered rock has been eroded by the rushing Colorado River that carved the mile-deep canyon almost 300 miles long. This can be seen on a much smaller scale in most local rivers and streams. In our backyard, for example, the tiny brook that runs through our property has moved its path at least 15 feet from west to east due to the erosion of one side of the brook and the deposition of soil and rocks on the other side. It is another of our everyday mysteries as to why this has happened and how far our brook will meander before it is finished remodeling our backyard nature trail.

Kerry and Mari were finding the results of weathering and erosion on Bare Mountain all the way from the bottom to the top. First they encountered the sand "beach" that filled Mari's shoes. High up on the mountain, rocks that contained silica or quartz had been weathered into the tiny sand particles and eroded away. Being the smallest particles, they were easy for the streams and rivulets created by rains to carry all the way to the bottom of the mountain.

As the girls climbed higher, they encountered the boulders that were weathered away from the original rock that composed the mountain. Gravity helped these boulders' erosion by tumbling them down to lower areas, where they became stepping-stones for people who climbed the mountain.

Farther up, Kerry and Mari encountered the talus (or scree) slope that delayed their hike. A talus slope is made up of small rocks that are constantly sliding down an incline that does not exceed 40 degrees. The top rocks are at the mercy of the weather and are broken into smaller pieces, but there is constant addition of rock to the slope. The newer rocks atop the others are always sliding downward, so they provide a very slippery surface to a climber. Mari and Kerry experienced this and so, I might add, did my wife, our dog, and I, as we tried to climb that same slope one day, years ago. It felt like walking up a slope of marbles. The force of friction is practically gone and the sliding rock offers little purchase for the hikers' feet.

Finally, as the children reached the top of the mountain where the weathering is continuing to produce the products witnessed before, the cracks and "doggy water bowls" are very prevalent. The water that collects in the depressions will expand and freeze in winter weather and, sooner or later, will crack a piece of the mountain into bits. Boulders will continue to break away. Someday, if the Earth does not produce another upheaval, the mountain will be as flat as the surrounding valley below—but not in our lifetimes.

In the early 20th century, German geologist Alfred Wegener suggested a theory of continental drift that explained the movements of the continents over long periods of time from one huge continent (named Pangaea) to what we see today. At that time, due to lack of sufficient evidence to support his theory, it was unacceptable to the majority of scientists. It was not until the 1950s that enough scientific evidence had been collected to allow his theory to have some credence and eventually, after much debate and investigation, lead to the accepted theory of today that explains how these plates move.

Scientists determined that Earth's crust was actually a set of between 7 to 12 plates floating on the surface of the Earth's mantle and constantly moving at the mind-boggling speed of 50–100 millimeters annually. These plates have enough force to cause earthquakes or to push against one another, causing tremendous pressures beneath the surface. This is what could have caused the fault-blocking that formed Bare Mountain.

Other mountains are made in different ways. Some mountains, like Mount St. Helens and Mount Rainier in Washington State, are formed and continue to be formed as *volcanic cones*. These mountains are easily identified by their cone shape. On May 18, 1980, Mount St. Helens erupted and launched 0.7 cubic miles of rock and debris into the air. It and 160 other active volcanoes are located in the Pacific Ring of Fire along the Western United States.

Other mountains are caused by opposite horizontal pressures within the Earth that push the areas between the pressures up into a folded position. Some of the remnants of these kinds of mountains can be seen along roadways and are identified by the upward or downward layers of the rock strata. These are known as *folded mountains*. Examples of this kind of mountain are the Himalayas in Asia. The Andes in South America are believed to have been formed when one plate actually was pushed under South America and thereby produced lots of volcanic activity. Because of this, scientists are more inclined to consider the Andes as a predominantly volcanic mountain range. These two examples also happen to be the youngest and tallest of the mountains in the world.

Another type of mountain is the *upwarped mountain*, caused by direct forces beneath the surface of the Earth, which push the rock above it upward, causing little deformation of the rock strata. Examples of this type of mountain are the Black Hills in South Dakota and the Adirondacks in New York. Walks up these kinds of mountains might result in discovering fossils at the top. The land at the summit of the mountain might have once been an ocean bed or prairie.

Mountain building and destruction have also created the phenomenon of the *monadnock*. Monadnocks stand as solitary entities, like lonely sentinels guarding the countryside. Monadnocks are made of volcanic rock that is resistant to erosion and weathering, and remain while the softer surrounding areas are eroded away, leaving it as a large rock remnant in a low valley. Mount Monadnock in southern New Hampshire, Mount Sugarloaf in Rio de Janeiro, and Stone Mountain in Georgia are examples.

A hiker climbing up any mountain, however it was formed, is likely to see the geological phenomena observed by Mari and Kerry when they climbed Bare Mountain. This is because all mountains undergo weathering and erosion over time. Even the tallest and most rugged peaks will someday become rounded and

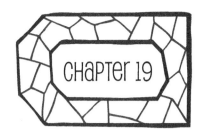

will have boulders, talus slopes, and sand on them. This cycle of mountain building and destruction goes on continuously on the Earth.

All of these theories about mountain formation and erosion are based on evidence obtained over hundreds of years. The processes are so slow that no one person or group can actually watch the total evolution of mountains. The plate tectonics theory is a wonderful example of how science works. Theories are proposed to explain certain phenomena, but until there is enough evidence to support the theories, they do not warrant the approval of the scientific community. Thus science knowledge grows by a constant search for data that will explain these phenomena. Theories are modified as new data are found until the scientific community is satisfied that the theory is sound. It is the function of theories to explain facts and observations. For example, the idea that change in living things had happened and was continuing to happen was evident. It was Charles Darwin who proposed the theory of natural selection as an explanation as to how this change came about.

reLaTeD IDeaS FrOM THe NaTIONaL SCIeNCe eDUCaTION STaNDarDS (NrC 1996)

K–4: *Changes in Earth and Sky*
- The surface of the Earth changes. Some changes are due to slow processes, such as weathering and erosion.

5–8: *Structure of the Earth System*
- Landforms are the result of a combination of constructive and destructive forces. Constructive forces include crustal deformation, volcanic eruption, and deposition of sediment, while destructive forces include weathering and erosion.
- Interactions among the solid Earth, the oceans, the atmosphere, and organisms have resulted in the ongoing evolution of the Earth system. We can observe some changes such as earthquakes and volcanic eruptions on a human timescale, but many processes such as mountain building and plate movements take place over hundreds of millions of years.

reLaTeD IDeaS FrOM BeNCHMarKS FOr SCIeNCe LITeraCY (aaaS 1993)

K–2: *Processes That Shape the Earth*
- Chunks of rocks come in many sizes and shapes, from boulders to grains of sand and even smaller.

3–5: Processes That Shape the Earth
- Waves, wind, water, and ice shape and reshape the Earth's land surface by eroding rock and soil in some areas and depositing them in other areas, sometimes in seasonal layers.
- Rock is composed of different combinations of minerals. Smaller rocks come from the breaking and weathering of bedrock and larger rocks. Soil is made partly from weathered rock, partly from plant remains—and also contains many living organisms.

5–8: Processes That Shape the Earth
- Some changes in the Earth's surface are abrupt (such as earthquakes and volcanic eruptions) while other changes happen very slowly (such as uplift and wearing down of mountains). The Earth's surface is shaped in part by the motion of water and wind over very long times, which acts to level mountain ranges.

USING THE STORY WITH GRADES K–4

Obviously, this story is best followed by a field trip to a mountain such as Bare Mountain. Okay, so you teach in South Florida where the tallest landmark is the local landfill, or in Delaware where your tallest peak is about 100 meters (328 feet). If you have a stream that you can visit, you can introduce your students to some of the basic ideas of weathering and erosion. Failing this, you can demonstrate some of the ideas in a sand or water table.

You might start with a question such as, "What do you think caused the differences in the kinds of materials Mari and Kerry found on their hike?" Student responses may include any of the following:

- The water took the sand down farthest because it is the smallest.
- Ice can break up rocks.
- Rocks on rocks can be slippery.
- People can break rocks with their feet.
- The tops of mountains are the hardest rock.
- Rocks can break rocks when they fall on each other.
- When big rocks fall, they can break other rocks.
- Some rocks are made up of sand.
- Water doesn't move rocks very far.

These statements can be changed into questions such as

- Does water carry sand farther than it does rocks?
- How far does water carry rocks?
- Does how far a rock is carried depend on the rock size?

186

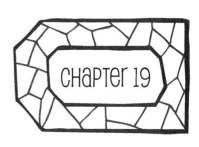

- Are some rocks made up of sand?
- Can big rocks break other rocks?
- Can ice break down a rock?

Visiting a stream and watching where the sand and rocks are distributed in the faster flowing areas can help children investigate some of these questions. Your students will see that the sand is much farther downstream than the rocks, and that these are sorted by size in the areas where the stream flow lessens. Barring this, you can place a mixture of sand and gravel in the school water or sand table or even on a pie tin, tilted so that you can pour water through the mixture to lower levels and see the difference in where the materials are deposited.

If you can find rocks that have cracks in them, you can place water in the cracks and then freeze them to see the effect of expanding ice on the rocks. Deeper cracks, obviously, would be most dramatic. Rubbing a piece of sandstone will produce a great deal of sand.

Considering the concept of time in the billions of years, it is best to wait until children are older to introduce it, as the idea of these large numbers is no more intelligible to young children than is the national debt to the average citizen. It is enough that the children are acquainted with the types of actions that can wear away those things that at first glance seem indestructible. A visit to a local graveyard may also be appropriate for older children in your age group. Please read the following section as well for ideas that may be appropriate or modifiable for your students.

I once visited a classroom where young children were introduced to time in a very exciting way. The teacher obtained a long piece of paper on which the students wrote something that was interesting about each day of the school year. She began on the first day of school and continued with a comment on each succeeding day until the end of the term. The students were amazed at how many things had happened and how very long the paper had to be to contain all of the happenings of their year. One hundred and eighty days does not seem so long to us, but seeing it as a physical display gave the students a new way of looking at the passing of time.

USING THE STORY WITH GRADES 5–8

Our teachers chose to begin with a probe titled "Mountain Age," from *Uncovering Student Ideas, Volume 1,* by Keeley, Eberle, and Farrin (2005) or "Mountain-top Fossil" from *Volume 2* of the same series (Keeley, Eberle, and Tugel 2007). The first probe asks students to share their ideas about the processes that affect the shape of mountains and to describe their thinking about mountain formation and evolution. It will give you a good idea about what your students know about weathering and erosion, and will provide a wonderful springboard into a discussion about the topic.

Another barrier to understanding the process of mountain building is grasping the concept of large amounts of time. Billions of years are difficult for students to comprehend, and most young children cannot fathom the span of time needed for geological changes to occur. Teachers who have dealt with this problem have had some success in setting up a yearlong timeline in the classroom with long pieces of calculator tape so that the students can see the length of one school year and the events that mark its passage. This helps them see the linearity of time and put an understandable period of time in perspective.

The "Mountaintop Fossil" probe asks students to explain how a fossil seashell could have found its way to the top of a mountain. We have found that some students think of a fossil as synonymous with a seashell and are unfamiliar with the commonly huge magnitude of the rocks in which the fossils are hidden. Once again, students do not have many direct experiences with the changes that go on in geology due to the fact that those changes take place over such long periods of time. But there are ways to simulate these changes and to find places where weathering and erosion happen in a time frame that is meaningful to a student of this age. I can also recommend the article "Chipping Away at the Rock Cycle," by Debi Molina-Walters and Jill Cox (2009) in *Science Scope*. It connects rock cycles with weathering and erosion and can be found online at *www.nsta.org*. In this article, the authors describe how to simulate the formation of rocks using cooking chips, heat, and pressure. If you are going to venture into the rock cycle, this is a must read.

A visit to a graveyard is a tried-and-tested field trip where, if there are older gravestones, the weathering of the markers is easy to observe. It also becomes obvious from looking at dates, that the newer ones are carved from granite and marble that resist weathering much more than the older limestone and sandstone markers of years ago. (It makes a great deal of compassionate sense to find out if any of your students have lost a family member lately and to talk to the parents about making the trip a pleasant one.) It may also be necessary to get permission to visit the graveyard and to talk to the students about being sensitive to the memories of the deceased.

I ask you to read both grade-level sections to see if any of the ideas match the needs and abilities of your students. If you are going to examine graveyard weathering, I offer two articles that are greatly informative and are easily available on the NSTA website. The first is "Cemeteries as Science Labs," by Linda Easley (2005) in *Science Scope*. This article offers wonderful suggestions, not only for a look at weathering through the decades, but also at how the topic can be integrated with mathematics, social studies, language arts, and archaeology. The other article is from the "Science Sampler" section of *Science Scope* and is written by the team of Robin Harris, Carolyn Wallace and Joseph Zawicki (2008). "Chemical Weathering: Where Did the Rocks Go?" helps you design an inquiry unit on weathering that shows the effects of various chemicals in your local community.

Visits to graveyards are often inexpensive or free if the site is a walkable distance from your school. It is a good idea to visit the site first and determine if it is old enough (from about 1850) and has been used during the past 50 years so

that differences between the harder headstones and softer headstones are obvious. As described in the Easley article, one can also find data to support theories about epidemics, childhood diseases, war, and other aspects of society, past and present. If you are the kind of teacher who likes to integrate your curricula, you cannot find a more suitable place to start.

You may also have in your school a stream table that will allow you to simulate the action of running water on various types of material. Students can develop hypotheses about what will happen if certain variables are changed in the use of the table and check out the results with near-instant feedback.

Lacking a stream table, a simple activity might be to put samples of various-size rocks and sand in a container of water. Children are asked to suggest what the container will look like tomorrow if it is shaken and allowed to stand overnight. The next day's observation will show that the various-size particles will layer out according to size. The densest will be at the bottom and the least dense at the top. Thus it shows that the sand will remain in running water longer and end up at the bottom of the mountain while the larger rocks will be dropped by the eroding stream sooner and closer to the top. Children can also simulate the formation of folded mountains using plasticine, and I direct you to *www.coaleducation.org/lessons/middle/mountain_building.htm* for directions.

reLaTeD BOOKS aND NSTa JOURNaL arTICLeS

Driver, R., A. Squires, P. Rushworth, and V. Wood-Robinson. 1994. *Making sense of secondary science: Research into children's ideas.* London and New York: Routledge Falmer.

Keeley, P. 2005. *Science curriculum topic study: Bridging the gap between standards and practice.* Thousand Oaks, CA: Corwin Press.

Keeley, P., F. Eberle, and C. Dorsey. 2008. *Uncovering student ideas in science: Another 25 formative assessment probes, volume 3.* Arlington, VA: NSTA Press.

Keeley, P., F. Eberle, and L. Farrin. 2005. *Uncovering student ideas in science: 25 formative assessment probes, volume 1.* Arlington, VA: NSTA Press.

Konicek-Moran, R. 2008. *Everyday science mysteries.* Arlington, VA: NSTA Press.

Konicek-Moran, R. 2009. *More everyday science mysteries.* Arlington, VA: NSTA Press.

Monnes, C. 2004. The strongest mountain. *Science and Children* 42 (2): 33–37.

reFerenCeS

American Association for the Advancement of Science (AAAS).1993. *Benchmarks for science literacy.* New York: Oxford University Press.

Easley, L. 2005. Cemeteries as science labs. *Science Scope* 29 (3): 28–31.

Harris, R., C. Wallace, and J. Zawicki. 2008. Chemical weathering: Where did the rocks go? *Science Scope* 32 (2): 51–53.

Keeley, P., F. Eberle, and L. Farrin. 2005. *Uncovering student ideas in science: 25 formative assessment probes, volume 1*. Arlington, VA: NSTA Press.

Keeley, P., F. Eberle, and J. Tugel. 2007. *Uncovering student ideas in science: 25 more formative assessment probes, volume 2*. Arlington, VA: NSTA Press.

Molina-Walters, D., and J. Cox. 2009. Chipping away at the rock cycle. *Science Scope* 32 (6): 66–68.

National Research Council (NRC). 1996. *National science education standards*. Washington, DC: National Academies Press.

CHAPTER 20

WHAT ARE THE CHANCES?

D ad looked up from the evening newspaper.
"Hey, listen to this, guys. It says here that a U.S. spy satellite is likely to break up this week and fall to Earth. Think we ought to get out our lead umbrellas?"

"Where is it coming from?" asked Sam, who was lying on the floor playing with his toy cars.

"From outer space," said Dad. "And it weighs a couple of tons, so that should make quite a dent in the sidewalk!"

"Oh, come on, George!" said Mom, who was watching TV. "You're going to scare Sam and me both. We won't go outside all next week if this is true."

"Well, it does seem to be a little dangerous, although I'll bet most of it burns up in the atmosphere before it hits the Earth."

"Do you think it will really hit here?" asked Sam. "That is really scary."

"Nah," said Mom. "Daddy is just trying to have some fun with us. It won't likely fall on us, will it, George?"

"Well, it does say that it weighs several tons so it all can't burn up, or can it? Anyway what are the chances it will land here in our town anyway?"

"Does the burning add to global warming, Dad?" asked Sam.

"Nah, that's a whole different story, Sam. These things burn up fast in the upper atmosphere and don't add a lot of heat."

"The Earth is pretty big, isn't it Dad?" asked Sam, who had stopped playing and was really looking worried.

"I'm sorry, Sam, I didn't mean to scare you. But there is a lot of space junk out there and sooner or later it will have to fall down because, as the old saying goes, "What goes up…"

"…must come down,'" Mom finished his sentence.

"Gosh, I hope it doesn't fall on our house," said Sam with a bit of fear in his voice.

"Look, like you said, the Earth is pretty big and chances are it won't fall anywhere near here," said Mom.

"But, it has to land somewhere," said Sam. "So why not here?"

"It all has to do with what the chances are," said Mom. "Let's look at the globe and see what we can find out."

That night Sam went to sleep feeling a lot safer. Why do you suppose that was the case?

PURPOSE

The Earth is at least 71% ocean, so that leaves only 29% for the rest of the planet, which also includes lakes, deserts, ponds, and habitable land. Students should be able to see how much of our planet is covered in water. Our students, and indeed most of our population, may not really be aware of how much of our Earth is covered in water. Some have suggested that we should be called "Oceana" rather than Earth because of this. This story will help introduce students not only to a clearer idea of the characteristics of our planet, but also to the concepts of probability, space exploration and satellites, space junk, and planetary responsibility. Because Sam asks if the burning object can add to global warming, you may segue into that topic if you wish.

RELATED CONCEPTS

- Atmosphere
- Probability
- Satellites
- Global warming

DON'T BE SURPRISED

Your students will probably not be aware of the amount of water that covers the Earth's surface. While students have all seen globes, understanding of the vast amount of oceans and water does not really kick in unless they have been alerted to look at the globe with this in mind. Any student that has crossed an ocean, either by air or boat, will be aware of how long it takes to cross these huge bodies of water, but it may have been just another "Are we there yet?" moment for them.

CONTENT BACKGROUND

In 325 BC in Alexandria, Egypt, Eratosthenes calculated the Earth's circumference at approximately 25,000 mi. Since the actual circumference is 23,902 mi (40,075 km), he wasn't far off, and given the crude measuring instruments of his day, his estimate is even more remarkable.

Scientists tell us that Earth's surface is about 510,000,000 km^2, of which approximately 361,000,000 km^2 (about 71%) is covered in water. For good reason, when viewed from space, Earth has been called the "Blue Planet." In addition, deserts cover about 20% of the Earth's surface and ice caps about 2%. When you add all this up, the populated area of the planet is only about 12% of the surface. This means that the probability of any space object landing in a populated area is around 1 in 10. Put another way, the odds are 10 to 1 *against* a space object landing in any populated area. And when you figure the small space your house takes up on this Earth, the odds are astronomical.

Of course, meteorites and space debris do land in populated areas because of the laws of probability. Probability only states that the likelihood of an event is high or low. It does not say that it is impossible. Still, Sam can sleep easy, knowing the low odds of being hit by space debris.

Most of the space materials that enter Earth's atmosphere hit the air and, because they are traveling so fast, cause friction that gives off energy. Most of this energy is either given off as heat energy or light energy. When spare materials do burn on entry, they are called *meteors*, and as we see them we often call their trails "shooting stars." If they are large enough to withstand the complete burning, they do hit the Earth and are called *meteorites*.

Over the millions of years that Earth has been bombarded by meteorites, some very large ones have left big craters, such as the Barringer crater in northern Arizona. It is 4,100 ft. in diameter (1.2 km) and 570 ft. deep (173 m). It is thought to have hit Earth between 20,000 and 50,000 years ago. Nothing in recent history has been noted that would compare. Most of the smaller debris interacts with the atmosphere of Earth and the friction between the debris and the atmosphere causes these smaller objects to burn up and never reach the ground.

If by any chance you decide to use this topic to segue into the area of global warming, I can recommend an article published in the journal *Science Scope* titled "Issues in Depth: Inside Global Warming" (Miller 2006). It gives a wonderful overview of the issues surrounding global warming for the teacher. It is available online from the journal archives at *www.nsta.org*.

related ideas from the national science education standards (nrc 1996)

K–4: *Properties of Earth Materials*
- Earth materials are solid rocks and soil, water, and the gases of the atmosphere.

5–8: *Structure of the Earth System*
- Water, which covers the majority of the Earth's surface, circulates through the crust, oceans, and atmosphere in what is known as the water cycle.

related ideas from benchmarks for science literacy (aaas 1993)

6–8: *The Earth*
- The Earth is mostly rock. Three-fourths of the Earth's surface is covered by a relatively thin layer of water (some of it frozen), and the entire planet is surrounded by a relatively thin layer of air.

USING THE STORY WITH GRADES K-4

In Keeley and Tugel's book, *Uncovering Student Ideas in Science, Volume 4* (2009), there is a probe called "Where Would It Fall?" Students are asked to predict where an object from space would land on the Earth. Some of the possible answers include the desert, populated areas, oceans, glaciers, the largest continent, and a body of freshwater. It would be interesting for you to give this probe to your students to see where they stand on this issue. The probe also asks them to explain their answers. As the story suggests, a look at the globe may help answer any questions they may have.

Some might say that mathematical probability has no place in the early grades, but experience tells me otherwise. I have had great success in as early as second-grade classrooms with an activity that helps children see the difference between possible and probable. I will describe this for you.

I distribute three fresh pea pods to each of the students and ask them to open the pods, count the number of peas in each pod, and record that number. Each child also has three small squares of paper, one representing each pod's number of peas. I draw a horizontal line on a large piece of paper and attach it to the board. On the horizontal line labeled "Number of Peas in a Pod," I place numerals from 0 to 10, equally spaced. I ask each student to come up and paste their square above the number that corresponds to the number of peas in each of his or her pods. This forms a histogram that normally will be a bell-shaped curve. In the middle will be the most often recorded number of peas. We discuss this and note how few there are at either end of the graph.

Then, having saved a dozen or so pods in a bag, I ask them to predict the number of peas in a pod that has yet to be opened. I then open them one by one, asking for predictions for each pod. Most students will pick the most common number, while an occasional child will pick a number at random with the explanation that it is his or her favorite number. I then place a different-color square above the number in the newly opened pod. Normally, most of the new pods will have a number that is in the center of the graph (the *probable*), but occasionally one or two will land on either end (the *possible*). Some students catch on quickly and predict that it will be between one of the three central numbers. By the time we finish with the pods in the bag, almost all of the students seem to understand the difference between probable and possible. Each time I reach into the bag I say, "Is it possible that there will be two peas in this pod?" (Yes). "Is it probable?" (No). This done over and over drives home the meanings of possible and probable.

Another game I picked up from the same probe material mentioned above is the passing of an inflatable globe from student to student at least 25 times or more. Gather the students in a circle and pass the inflated globe back and forth among the students. When the globe is caught by a student, that student sees where his or her hands are placed on the globe, for example, desert, populated land, or ocean. The group records this, and this record will show that the majority of the time a child's hands were on ocean when the globe was caught.

This may also be a great time to do a little integrating with geography, which has been so neglected in our curriculum lately. Continents, countries, and even major lakes can be pointed out while playing with the globe.

USING THE STORY WITH GRADES 5-8

Middle school children will have a better understanding of the sizes of oceans on Earth's surface but may still be amazed by the amount of ocean water that exists. The activities mentioned in the K–4 section are equally appropriate for this age group. It never hurts to reinforce the ideas of probability and possibility, and the globe passing activity will help you set the stage for some more sophisticated work. You may, for example, want to introduce students to the age-appropriate means of calculating the area of a sphere and develop estimations of surface water and habitable land to help them see the importance of the integration of math and science.

Surface areas of spheres are calculated by using the formula $S = 4\pi r^2$ where r is the radius of the Earth and pi (π) is approximately 22/7 if you are working with fractions and 3.1416 if you are working with decimals. The surface area of the Earth is approximately 196,940,400 mi^2 (510,065,600 km^2) and the mean diameter is 7,913 mi (12,735 km). The students will be able to reach a reasonable approximation of Earth's surface area and will be interested to find out that only 29% is land. Further investigation will allow them to see that of that 29%, deserts, plateaus, ice, and mountain ranges are uninhabitable, lessening the chances of a small satellite landing on a home.

Students are probably aware of the dangers of the space shuttle entering the Earth's atmosphere and how the shuttle is protected by heat shields. This allows the students to inquire about the phenomena of the meteor showers that occur at various times of the year when the particles from comet tails enter Earth's atmosphere. Meteor showers occur regularly and are usually advertised in the media. Particles enter the atmosphere at tremendous speeds and the friction with the atmosphere leads to their incineration. The result is a streak of light crossing the night sky.

Students in this age group may bring up the issue of global warming and the controversy that surrounds it. I am not sure why, but they often do. Perhaps it is because of the topic of the Earth brings up questions that concern them. I feel obliged to offer some suggestions here just in case your class brings up the subject. There are ample data showing that the Earth's surface temperature has been warming over the past 250 years. The controversy is based on some scientists who say that it is not *human* activity that has produced enough of the greenhouse gases to account for the change and that the change is due to a normal cycle of temperature fluctuations. These skeptics should be able to explain how the addition of so many greenhouse gases could *not* be affecting the climate. They cannot explain why the warming trend is increasing at such a great rate.

Some say that the whole idea of global warming is a political maneuver and that Earth's warming trend is a natural phenomenon that repeats itself over the eons. And, it has—but these fluctuations were before humans began to produce carbon footprints. This alone should spark some interest in how we are adding to whatever causes cyclical temperature fluctuations.

This allows an opportunity for classroom debates to be planned on the topic of global warming and the assertion that it is being caused by human activity. Critical thinking and analysis of various publications can be used to try to reach a conclusion. In our experience, the debate is a wonderful way for students to take a stand and then provide evidence to back up their claims. After all, scientists have access to the same data, but interpretation of these data is not always consistent.

This, of course, is one of the major factors relevant in the history of science. Implicit in the use of data is sorting out fact from fiction, relevant data from irrelevant, and biased from unbiased reporting. This is excellent practice for students engaged in drawing conclusions from a database that is growing by leaps and bounds due to the increase of technological devices available. I recommend that you obtain a copy of "Information Literacy for Science Education: Evaluating Web-Based Materials for Socioscientific Issues" by Klosterman and Sadker (2008) from the *Science Scope* archives (*www.nsta.org*). Do not be frightened by the title of this article! It gives very practical tips about how to help students sort through the plethora of information available to them. It will help you not only with this topic but with any socioscientific issue you may have to deal with.

RELATED BOOKS AND NSTA JOURNAL ARTICLES

Driver, R., A. Squires, P. Rushworth, and V. Wood-Robinson. 1994. *Making sense of secondary science: Research into children's ideas.* London and New York: Routledge Falmer.

Keeley, P. 2005. *Science curriculum topic study: Bridging the gap between standards and practice.* Thousand Oaks, CA: Corwin Press.

Keeley, P., F. Eberle, and C. Dorsey. 2008. *Uncovering student ideas in science: Another 25 formative assessment probes, volume 3.* Arlington, VA: NSTA Press.

Keeley, P., F. Eberle, and L. Farrin. 2005. *Uncovering student ideas in science: 25 formative assessment probes, volume 1.* Arlington, VA: NSTA Press.

Keeley, P., F. Eberle, and J. Tugel. 2007. *Uncovering student ideas in science: 25 more formative assessment probes, volume 2.* Arlington, VA: NSTA Press.

Konicek-Moran, R. 2008. *Everyday science mysteries.* Arlington, VA: NSTA Press.

Konicek-Moran, R. 2009. *More everyday science mysteries.* Arlington, VA: NSTA Press.

references

American Association for the Advancement of Science (AAAS).1993. *Benchmarks for science literacy.* New York: Oxford University Press.

Keeley, P., F. Eberle, and L. Farrin. 2005. *Uncovering student ideas in science: 25 formative assessment probes, volume 1.* Arlington, VA: NSTA Press.

Keeley, P., and J. Tugel. 2009. *Uncovering student ideas in science: 25 new formative assessment probes, volume 4.* Arlington, VA: NSTA Press.

Klosterman, M., and T. Sadker. 2008. Information literacy for science education: Evaluating web-based materials for socioscientific issues. *Science Scope* 31 (8): 62–65.

Miller, R. 2006. Issues in depth: Inside global warming. *Science Scope* 30 (2): 56–60.

National Research Council (NRC). 1996. *National science education standards.* Washington, DC: National Academies Press.

CHAPTER 21

HERE'S THE CRUSHER

Chores! My turn to do dishes again tonight," thought Eric. "Maybe if I have a second piece of pie, they'll forget."

"May I please have another piece of that super delicious pie?" Eric asked. "I think it's the best you've ever made," he said, hoping that flattering his sister would make her forget it was his turn at the dishes.

"Oh certainly, little brother, and since you are doing dishes tonight, I suppose you want to use the same plate you're using now." Janny smiled knowingly at him.

"Rats, she remembered!" Eric decided that he'd eat the pie anyway and then tackle the darn dishes.

He finished the pie, downed a glass of milk, and headed for the sink and the dinner dishes, cleared from the table and ready for his tender loving care. He filled the tub with soapy water, began to wipe and rinse the dishes, and put them into the drainer. When he had finished, he spotted a plastic soda bottle left on the kitchen counter waiting to go into the recycle bin.

"Guess I might as well rinse it out too, even though it isn't a dinner dish," he thought. Eric thoroughly rinsed the bottle out with very hot water, poured out the water and screwed the cap back on. He placed it back on the counter and started to leave the kitchen. Suddenly, he heard a crackle behind him and turned around just in time to see and hear the soda bottle as it began to collapse into itself. It crackled and crushed as though someone were squeezing it.

Eric opened the cap and the bottle returned to its original shape. He repeated the rinsing and capping process again and again and marveled over the result each time.

"Hey, guys," he shouted. "Come look at this."

Everyone had an opinion about it.

Big sister Janny said Eric had caused a vacuum when he poured the water out because the air in the bottle went out with the water, but couldn't explain why the crushing didn't happen immediately. Mom thought it was a matter of the kind of plastic from which the bottle was made and that it shrunk when heated by the hot water. Dad agreed.

Other members of the family thought it happened because of the water being poured out in the rinsing, while others wondered if the container had to be plastic.

There were a lot of thoughts and lots of wonderings. Meanwhile Eric began thinking in terms of "what ifs…" and began to try a lot of his ideas right there at the kitchen sink. The idea of spending time at the sink was no longer a chore but fun!

PURPOSE

Anyone who has rinsed out a plastic soda bottle may have had this experience. But I wonder how many have noticed it and had some of their own "what ifs…." That is the purpose of this story. Let's explore air pressure and its importance in our lives as an everyday science mystery.

RELATED CONCEPTS

- Air pressure
- Expansion and contraction
- Temperature
- Vacuum
- Heat energy

DON'T BE SURPRISED

Even though every weather reporter talks about high- and low-pressure areas, most of these comments go over our heads, or as the saying goes, "in one ear and out the other." Many of your younger students do not believe that air around us has mass or weight, let alone exerts pressure on us and on everything around us.

Many students believe that air, or any gas, is in the same category as abstractions such as thoughts. If you have viewed any of the *A Private Universe* films, you may remember Jon, the middle-school student who absolutely refused to believe that air (or any gas) took up space unless it was moving and was called "wind" or was in the form of dry ice. Most children are not aware that "flat" soda weighs less than fresh soda.

One major preconception that can be expected is that children think of air and oxygen as synonymous. Air is of course made up of many gases of which oxygen is only one. Oxygen amounts to about 21% of the total chemical makeup of air.

So, the idea that the atmosphere in which we walk actually has mass and can exert pressure on our world may seem completely ridiculous to many of your students. Of course they won't tell you that if asked directly. The story itself, however, is a type of formative assessment, and a discussion about the various opinions given will give you valuable information on what the students believe about the air that surrounds them.

CONTENT BACKGROUND

Above us, below us, inside us, and all around us lies a mixture of gases we call our atmosphere. This atmosphere, our *air*, is made up of molecules of gases that take up space and therefore have mass. We take it in and release it from our bodies, as do all animals. It is vital to the life of plants because it provides the carbon dioxide from which they take the carbon to build their cells using photosynthesis. Air contains

water vapor and myriad minerals and particles that float around in it, including pollutants. When it moves, we feel it as wind. When it is absent, we cannot breathe.

Air exerts pressure on everything that exists on this Earth. At sea level the pressure is about 14.7 pounds per square inch (psi). The amount of pressure decreases as we rise above the Earth's surface. Technically the atmosphere ends at about 120 km (75 mi.) away from the Earth. Five different levels have been delineated in the atmosphere, each indicating a decrease in pressure with altitude.

At the top of Mount Everest at 8,850 m (29,000 ft.), the pressure is about 50% of that at sea level. Commercial airplanes fly at around that altitude, at an average of 33,000 ft. When you fly in a commercial airplane, the cabin in which you sit is pressurized to what the atmosphere would be at somewhere between 6,000 to 8,000 ft. (1,830 to 2,440 m). Although this is not sea level, the difference does not usually cause problems for most people. Airplanes must be pressurized for two reasons. First, the shape of the plane would change at higher levels and possibly cause structural damage. Second, passengers might become ill at the very low pressures found at high altitudes. The oxygen in the airplane cabins is also reduced at higher altitudes, but not usually to the point where you notice it unless you have a medical problem that requires oxygen to be at a sea level standard.

Pressure changes can be uncomfortable. You may have experienced discomfort when an airplane you are riding in ascends or descends, and your ears feel the pressure change. Air has been trapped in your ear canal, and as the plane changes altitude, you notice the difference between air inside your ear and out. The eardrum can retract to the point where some people actually feel pain. Others are merely inconvenienced until the pressure equalizes and their ears "pop." You may also feel this when descending quickly from a high mountain in your car or traveling in a fast elevator.

The envelope of air that is the atmosphere serves a very important function for life on Earth. It filters out a great deal of ultraviolet light and provides a shield for escaping radiation, therefore keeping the planet warm and reducing temperature extremes between day and night. This retention of heat energy that modifies the temperature of the Earth is known as the greenhouse effect. Scientists have ascribed the recent rapid increase in global warming to gases from burning fuel that have formed another layer in the atmosphere that prevents heat energy from escaping in a normal way.

Atmospheric pressure is measured as the downward pressure of the weight of air above the place where it is measured. As we mentioned before, it is greatest at sea level and less at higher altitudes. Air pressure is also affected by temperature and the amount of water vapor in the air. High-pressure air is usually more dense and dry, while low-pressure air is usually less dense and moist. This may seem counterintuitive, but the water vapor molecule is lighter than the average molecules that make up the air. Therefore, the average density (which is the mass divided by the volume) of the air is less and less pressure is exerted. This is called a "low." Conversely, a mass of dry air is denser and therefore exerts more pressure and is called a "high."

Dry air is usually cooler than moist air. So when cool air moves into your area, you can usually expect the cooler, dryer, and clearer conditions that go with high pressure weather. The opposite is true of the masses of warm, moist air.

Most of us in our years of schooling have seen the demonstration of the can of water heated so that the water boils and the vapor clouds are seen emerging from the can. The teacher placed the lid on the can, and as we watched in amazement, the can seemed to be crushed by an invisible hand. We were told that it was the air pressure that crushed the can and that the boiling of the water took the air out of the can so that it had no resistance to the air pressure around us. Did we believe it? I remember seeing it at least a half dozen times over my years in school, and by the time I was in high school physics I think I finally believed it and perhaps really began to understand it. I never got to touch the apparatus nor did I have the opportunity or equipment to try it at home.

Now with the advent of plastic bottles, our students can play safely with this activity as long as they want. Homework? Why not—as long as parents know what is going on and will supervise so that kids don't try boiling water. The warm water expands the air inside the bottle, which causes a lot of that air to leave through the open top. When you screw the cap back on after the bottle is empty, there is less air inside and it is warm. As it cools, it takes up even less space so the room temperature air pressure pushes on the outside of the bottle and crushes it. You can make it happen even faster if you pour cold water over the bottle after it is sealed. The air inside will cool faster and the reaction will be accelerated.

From personal experience, I can vouch for the tremendous pressure of the atmosphere. I was sold a new gas cap for my car that turned out to be nonvented. As I drove, gasoline was taken from the gas tank but no air was allowed to replace it due to the nonvented cap. The gas tank of the car collapsed from the pressure of the atmosphere! Fortunately, I was awarded a new tank and a new gas cap. The important thing here is that the gas tank was not a flimsy piece of metal but a substantial structure. However, with air pressure at 14.7 pounds per square inch, it was no match for the atmosphere after its contents were emptied, causing it to become a virtual vacuum. Like the plastic bottle, it was crushed from the outside!

related Ideas From The NaTIONaL SCIENCE EDUCaTION STaNDarDS (Nrc 1996)

K–4: Changes in Environments

- Changes in environments can be natural or influenced by humans. Some changes are good, some are bad, and some are neither good nor bad.

5–8: Structure of the Earth System

- The atmosphere is a mixture of nitrogen, oxygen, and trace gases that include water vapor. The atmosphere has different properties at different elevations.

RELATED IDEAS FROM BENCHMARKS FOR SCIENCE LITERACY (AAAS 1993)

K–2: Energy Transformations
- The Sun warms the land, air, and sky.

3–5: The Earth
- Air is a substance that surrounds us and takes up space. It is also a substance whose movements we feel as wind.

6–8: Processes That Shape the Earth
- Human activities such as reducing the amount of forest cover, increasing the amount and variety of chemicals released into the atmosphere, and intensive farming have changed the Earth's land, oceans, and atmosphere.

USING THE STORY WITH GRADES K–4

I like to start by asking the students if they have done dishes and have had the experience described in the story. Surprisingly, some have and are eager to tell you about it. Obviously, those who have not seen this firsthand will want to try it. I suggest that you demonstrate it for them by having them tell you what to do at each step. These steps can be written down on a large sheet so they can be analyzed in an effort to understand what might have happened.

It is also a good idea to have a half dozen identical bottles so when the children suggest different changes in the procedure, the bottles can be compared to one another. For example, if the children ask if warmer water makes a difference in how much the bottle is crushed (and they usually do ask that), you can try this and see.

If you have a sink in your room with hot and cold water, it will be easier for you to demonstrate this phenomenon. If not, you will have to have several containers of different temperature water at your disposal or else a hot plate to warm the water. It is also a good time to introduce your students to a thermometer if you have not already done so. This way, there will be data to record and analyze. Variables may include

- Temperature of the water
- Length of time you leave the water in the bottle
- Size of the bottle
- How much water you put in the bottle

- How the water is emptied from the bottle
- How the bottle is cooled
- Shape of the bottle
- The thickness of the bottle's plastic
- Difference in procedure, such as just heating the bottle from the outside

Children will be aware of procedure. You may put water in and swirl it around, or may pour it out quickly or gradually. Ask your students if it makes a difference if you follow the same procedure exactly each time. They are usually sticklers for following the exact routine.

Even though Eric did not cool the bottle quickly, you may want to add this to your procedure and ask the children to predict if the shrinking of the bottle will change in any way. This way, you introduce the idea of cooling as an important part of the solution to the problem. You probably will have to give hints as to the source of the pressure even though the children may still have trouble with this concept at the early elementary stage. Still, it gives them something to ponder over time and will add one more experience to those that eventually lead them to accept the scientific view of air pressure.

If you are brave, you may set up a center and allow students to try the investigation themselves. Or as an alternative, you may want to involve parents by sending a note home explaining what the students are supposed to do and asking that they supervise so that the children do not use any dangerous procedures. Warm water from the normal hot water tap is usually enough to give a reasonable reaction.

USING THE STORY WITH GRADES 5–8

Older children have probably had some experience like Eric's and will have opinions on why the bottle was crushed. Starting out with a "Best Thinking" sheet is a good way to begin the thinking process. Rest assured that your students will want to see this phenomenon and you can either proceed as suggested in the section above or, if your facilities allow, let the students test their ideas themselves. If you develop a list of possible variables as suggested previously, you can ask groups of students to try each of the various tests and to report on their findings. Predictions are important here, as always, and having students record their findings in their science notebooks will allow a good discussion after the lab work is completed.

I cannot overstress the importance of class discussion of this topic. Your leadership of the discussion is very important as you listen to what your students are saying and respond to them in as conversational a way as possible. The more you can get them to interact with one another, and not through you, the better the discussion will be. Dialogue among the students will bring out a great many ideas, and the arguments will allow the students to have their say and then have the opportunity to revise their thinking on the basis of what they have heard or said. It is always a good idea to have the materials available for students to demonstrate their points as needed.

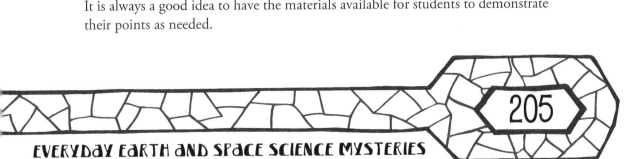

Of course, doing it at home is another alternative—with the caution that parental supervision is important, even more so with children of this age because of their fearlessness and frequent lack of judgment about safety.

Lastly, if you have access to the NSTA website, you can download the article "Torricelli, Pascal, and PVC Pipe" in the *Science Scope* archives (Peck 2006). In this article, the author has some great ideas about using straws, tubing, and PVC pipe to measure atmospheric pressure. I recommend it highly. The use of straws is a wonderful way to expound on one of the most popular misconceptions, that of our *sucking* up drinks through a straw. Actually, we lower the pressure in the straw and the atmospheric pressure *pushes* the liquid up into our mouths!

related BOOKS and NSTA JOURNAL articles

Driver, R., A. Squires, P. Rushworth, and V. Wood-Robinson. 1994. *Making sense of secondary science: Research into children's ideas.* London and New York: Routledge Falmer.

Keeley, P. 2005. *Science curriculum topic study: Bridging the gap between standards and practice.* Thousand Oaks, CA: Corwin Press.

Keeley, P., F. Eberle, and C. Dorsey. 2008. *Uncovering student ideas in science: Another 25 formative assessment probes, volume 3.* Arlington, VA: NSTA Press.

Keeley, P., F. Eberle, and L. Farrin. 2005. *Uncovering student ideas in science: 25 formative assessment probes, volume 1.* Arlington, VA: NSTA Press.

Keeley, P., F. Eberle, and J. Tugel. 2007. *Uncovering student ideas in science: 25 more formative assessment probes, volume 2.* Arlington, VA: NSTA Press.

Konicek-Moran, R. 2008. *Everyday science mysteries.* Arlington, VA: NSTA Press.

Konicek-Moran, R. 2009. *More everyday science mysteries.* Arlington, VA: NSTA Press.

references

American Association for the Advancement of Science (AAAS).1993. *Benchmarks for science literacy.* New York: Oxford University Press.

National Research Council (NRC). 1996. *National science education standards.* Washington, DC: National Academies Press.

Peck, J. F. 2006. Science Sampler: Torricelli, Pascal, and PVC pipe. *Science Scope* 29 (6): 43–44.

CHAPTER 22
ROTTEN APPLES

Ted and Steve were on their way home from school one October day and decided to take a shortcut, as they often did, through the old apple orchard near their homes. "You know, Steve, all of our homes are built on land that used to be part of a huge apple orchard. The owner sold off a lot of his old orchard to developers who built these homes. My mom remembers when all of this land, including our school, was apple orchard."

"Yeah, I remember hearing about that too," said Steve.

"I wonder what happened to all of the old apples that were all over this field and around our homes since the trees were cut down," said Ted.

"And look around this orchard now," said Steve, "There are all kinds of apples lying on the ground that didn't get picked or just fell after the harvest."

"You'd think that the apples from all of the years that trees have been dropping them would be around here and we'd be up to our knees in old apples," laughed Ted. "But actually, I wonder where they all go to. By next spring the ground will be clear and all of these apples will be gone. Does someone come out here and clean them all up?"

"I don't think so, unless they do that to make apple cider," replied Steve. "Yeah, that's probably what happens, or else in the spring we'd be walking over a lot of apples."

It was a nice warm fall day and the boys took their time walking home and stopped to look at some of the apples left on the ground.

"Man, look at these apples," said Ted, "They look half-rotten already. They wouldn't make cider out of these would they? They're all goopy and look like they have worms or something in them. They're not good for anything any more."

"I'll bet animals eat a lot of them but not all of them, 'cause there are still a lot of them left," offered Steve. "So why aren't they here in the spring? And where are the apples around our school and our houses?"

"I heard that they turn into soil," said Ted.

"Just like that? Magic?" asked Steve. "How can that be, soil is soil and it's always here, with or without apples. Soil is dirt, right, so apples can't turn into dirt! It has to be more complicated than that."

"I know," said Ted, "Let's take some of them home and put them in my yard and see what happens. We'll put them outside where the dogs can't get at them and we'll keep an eye on them. Nothin' like good ole observation, like Ms. Green keeps telling us in science class."

"Okay, as long as it's in your yard," said Steve, "I don't think my folks will like rotten apples in my yard and anyway our dog would eat them for sure. She eats anything, and I mean *anything*!"

And so the boys took a bunch of apples from the ground and took them home for some "good ole observation," just like Ms. Green kept telling them. And next spring...?

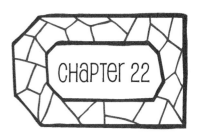
PURPOSE

In 1991 and 1992, John Leach, Bonnie Shapiro, and I did a study in which we interviewed approximately 400 students from the United Kingdom, Canada, and the United States about their beliefs surrounding the decay of an apple over a year's time. We found that children in all three countries had little understanding of the process of decay and almost totally ignored the role of microorganisms in the process. They believed that apples or any other previously living thing left on the ground miraculously turned to soil, were eaten by animals, or merely disappeared. Therefore the main purpose of this story is to help students explore how decay breaks down organic material so that it can be recycled within its ecosystem.

DON'T BE SURPRISED

While interviewing a high school student who had just finished Biology I for the 1992 study (Leach, Konicek, and Shapiro), I asked her if she thought that the materials from a rotting apple could possibly be used by another plant for its nutrients. She looked me as though I was completely addled. Then she said, "Wait a minute! Water recycles…! Ah, but not real stuff like apples."

This is typical of children (and many adults) who believe that matter just disappears into the soil or becomes soil. The idea of decaying material becoming part of the nonliving environment is completely foreign. Because many young (and older!) people think that bacteria are simply disease-producing organisms, they suppose that ridding the world of all bacteria would be beneficial. They lack knowledge of decomposers in the soil that help break down dead organic material. They probably do not understand the cycle of conservation of matter and recycling of materials into new growth. With the recent media buzz about composting and environmental concerns, children may have become more aware of recycling, but don't be surprised if they are unaware of how the process takes place.

Also, some of your students might believe that the Earth is getting heavier each year because of the leaves that fall and fruits of the field that are left behind. Of course, the gist of the story you have just read, put simply, is that the planet reclaims its own material, and uses it over and over again.

RELATED CONCEPTS

- Decay
- Decomposers
- Oxidation
- Decomposition
- Fungi

CONTENT Background

July 5, 2008: Today, in the cool of the morning, we picked up a truckload of bark mulch for use in our garden and watched as vapor was emitted from the huge pile each time the front loader scooped up a load to deposit in the bed of our truck. Inside that pile of bark mulch, tiny microbes were busy breaking down the organic material into its component parts: molecules of carbon and nitrogen and phosphorus-based compounds; and in the process, oxygen was being utilized and the oxidation process was releasing the energy stored in the plant material as heat. As the moist vapor condensed in the cooler air, little clouds appeared. I buried my hand in the bark mulch and could feel the heat, which I judged might have been as warm as 60°C. Indeed, in the field behind our home, the nearby college agriculture department often dumps their composting pile, and in the wintertime when the air is really cool, the clouds of vapor rising from it look almost like a steam vent.

What a wonderful process this is! Decomposition of organic matter by the microbes and fungi in the air and soil takes care of all of our unwanted organic material and returns the building blocks of their bodies to the earth to be used again and again. The same thing happens in the compost pile in our garden. We feed our vegetable garbage to the microbes and fungi, and in return, they give us beautiful, black, fertile soil to use in our garden. What a relationship we have with our tiny friends! We each have our jobs to do, by which we both benefit and, in turn, create a healthier planet.

In our compost pile each summer, we say hello to hundreds of worms, pill bugs, centipedes, and millipedes as they wiggle away from our intruding pitchfork while we turn their dinner table over to provide more oxygen and water. These creatures too are important, although they are not technically decomposers. Scientists call them *detritivores* or sometimes just scavengers. They do not completely digest all of the potential nutrients nor do they release all of the energy. The worms and other animals eat the garbage we put in for them and perform the first part of the decomposition process by transforming the large pieces of grass clippings, lettuce, avocado pits, and other plant material into more "bite-sized" pieces for the real decomposers, the microbes and fungi. By passing the unused portion of their meals through their bodies in a partially digested form, they speed up the process of composting. They also aerate the pile so oxygen and water can circulate and do their part in the process.

Ted and Steve will witness this phenomenon as they watch the apples over the fall and winter. The cold temperatures may slow the process down a bit, but sooner or later, the apples will become smaller and smaller as the decomposers use the food in the apples to satisfy their needs and release the essential nutrients back into the soil.

According to the U.S. Environmental Protection Agency nearly 25% of the solid waste that goes into landfills is made up of our organic garbage and lawn clippings. Almost all of this could be composted into usable material. The EPA website provides lists of desirable and undesirable materials for use in composting (*www.epa.gov/epawaste/conserve/rrr/composting/index.htm*).

related Ideas From the National Science Education Standards (NrC 1996)

K–4: *The Characteristics of Organisms*
- Organisms have basic needs.

K–4: *Organisms and Environments*
- All organisms cause changes in the environments where they live. Some of these changes are detrimental to the organism or to other organisms, whereas others are beneficial.

5–8: *Populations and Ecosystems*
- Decomposers, primarily bacteria and fungi, are consumers that use waste materials and dead organisms for food.

related Ideas From Benchmarks For Science Literacy (aaas 1993)

K–2: *Flow of Matter and Energy*
- Many materials can be recycled and used again, sometimes in different forms.

K–2: *Constancy and Change*
- Things change in some ways and stay the same in some ways.

3–5: *Interdependence of Life*
- Insects and various other organisms depend on dead plant and animal material for food.
- Most microorganisms do not cause disease and many are beneficial.

3–5: *Flow of Matter and Energy*
- Some source of energy is needed for all organisms to stay alive and grow.
- Over the whole Earth, organisms are growing, dying, and decaying and new organisms are being produced by the old ones.

6–8: *Interdependence of Life*

- Two types of organisms may interact with one another in several ways: They may be in a producer-consumer, predator-prey, or parasite-host relationship. Or, one organism may scavenge or decompose another.

6–8: *Flow of Matter and Energy*

- Food provides molecules that serve as fuel and building material for all organisms.

USING THE STORY WITH GRADES K–4

It might be a good idea before you read the story to your class to give them the probe "Rotting Apple" in Keeley and colleagues' *Uncovering Student Ideas in Science Volume 3: Another 25 Formative Assessment Probes* (2008). In this probe, four friends argue about why an apple, over time, disappears from view. Students have to choose from one of the six possible arguments. For young children, you may want to narrow the number of arguments so as not to confuse them. Regardless, you will probably find out, as we did in our 1992 study, that children do not consider small organisms capable of using the energy in the apple to break it down. Further, your students will not be aware of the particulate nature of matter and so do not think of the apple as being composed of particles so small that they can be absorbed by the soil and used again. But they may think about the possibility of small organisms using the apple for nourishment, and this is a first step in recognizing the importance of life forms smaller than the worms and insects that they may have seen eating rotting apples. Creating the "Best Thinking" chart to ask for their expectations about the future of an apple placed in a transparent container with some soil will also give you some idea as to what your children are expecting.

Cover the container so that mold spores do not escape into the classroom atmosphere. The children will see the apple begin to rot and observe the mold growing on the apple as it slowly shrinks into mush. It is a good idea to break the skin slightly so that the soil organisms can gain entrance. A bruised apple will also lend its own enzymes to the process. Decay would happen regardless, but bruising or scoring the fruit will speed up the process for impatient youngsters. Drawings and notes should be recorded in student science notebooks and a summary put on the class chart so that all can see a daily or weekly record.

In the end, the apple will seem to dissolve into the soil with possibly a piece of skin and a stem left behind. Since the container was covered, the students think that the apple has become part of the soil. They may also have never noticed the mold that grows and although they cannot see the bacteria, you could suggest to them that there are other organisms too small to see that are working along with the mold.

USING THE STORY WITH GRADES 5–8

Use the probe mentioned in the section above as it is written. The students will probably suggest that they be allowed to do the same thing that Steve and Ted do in the story and follow it up with some "good ole observation." You might consider asking your students to suggest variables in their setups and have several stations for observing apples. These might include the following:

- Types of apples
- Sizes of apples
- Apples with bruises
- Cut apples
- Apples in pieces
- Apples without soil in the container
- Apples with soil in the container
- Soils from various locations
- Different fruits such as bananas, grapes, kiwi, citrus.
- Comparison with leaves, grass clippings, and other common vegetation.
- Different temperatures
- Dry versus moist environments

Predictions can be listed at each of the various stations as to what the students think will happen in each case. Students will have to decide if the apples they use are to be gleaned from the ground or bought from a local orchard or grocery store. They should be concerned with keeping the variables at a minimum whenever possible. The website *www.break.com/index/fruit-decomposition-time-lapse.html* has a 30-second time-lapse video of two months of decomposition of fruit. Perhaps with today's ubiquitous technology, you and your students can set up your own time-lapse video of your decomposing apples. In case you might want to use the ever-popular bottle biology methods of investigating decomposition, you can visit their website at *www.bottlebiology.org*.

Your students may wonder how people help vegetables and fruits stay edible before the advent of refrigeration. The root cellar in most 17th- and 18th-century homes was used to store potatoes, carrots, and apples for use during the whole year. How did that work? Humidity and temperature were certainly factors but were there others? How did people keep the vegetables from freezing during the cold winters or rotting during warmer winters in warmer parts of the country? The reverse of learning about decomposition is learning how to preserve.

related Books and NSta Journal articles

Driver, R., A. Squires, P. Rushworth, and V. Wood-Robinson. 1994. *Making sense of secondary science: Research into children's ideas.* London and New York: Routledge-Falmer.

Keeley, P. 2005. *Science curriculum topic study: Bridging the gap between standards and practice.* Thousand Oaks, CA: Corwin Press.

Keeley, P., F. Eberle, and C. Dorsey. 2008. *Uncovering student ideas in science: Another 25 formative assessment probes,* volume 3. Arlington, VA: NSTA Press.

Keeley, P., F. Eberle, and L. Farrin. 2005. *Uncovering student ideas in science: 25 formative assessment probes,* volume 1. Arlington, VA: NSTA Press.

Keeley, P., F. Eberle, and J. Tugel. 2007. *Uncovering student ideas in science: 25 more formative assessment probes,* volume 2. Arlington, VA: NSTA Press.

references

American Association for the Advancement of Science (AAAS). 1993. *Benchmarks for science literacy.* New York: Oxford University Press.

Bottle Biology. *www.bottlebiology.org.*

Hazen, R., and J. Trefil. 1991. *Science matters: Achieving scientific literacy.* New York: Anchor Books.

Keeley, P., F. Eberle, and C. Dorsey. 2008. *Uncovering student ideas in science: Another 25 formative assessment probes,* volume 3. Arlington, VA: NSTA Press.

Leach, J. T., R. D. Konicek, and B. L. Shapiro. 1992. The ideas used by British and North American school children to interpret the phenomenon of decay: A cross-cultural study. Paper presented to the Annual Meeting of the American Educational Research Association. San Francisco.

National Research Council (NRC). 1996. *National science education standards.* Washington, DC: National Academies Press.

CHAPTER 23

IS THE EARTH GETTING HEAVIER?

Tom looked up from raking leaves and said to his cousin, Laura, "I think the Earth is getting heavier. Look at all of these leaves around us. There are more each year, and they weigh something, don't they?"

Laura leaned on her rake and looked at Tom with doubt. "Where did you get that brilliant idea?" she asked. "Is it because you hate raking so much that you think we just ought to let them sit there and add more weight to the Earth?"

"Well, it makes sense doesn't it? Each year the tree makes lots of leaves and each fall, they fall down and lie on the ground. Look over there in the woods; there are tons of leaves from last year and underneath those leaves are the ones from the year before. They have to add up to something."

"Well, I guess so and this has been going on for millions of years so the old planet must be getting really fat by now. No wonder the scientists say the Earth is slowing down," said Laura with a smile on her face.

I guess she doesn't really believe me, thought Tom.

"Okay," he said, "Let's go into the woods and take a look at the leaves on the ground and I'll show you what I mean."

The two children walked over to the woods and began sifting through the layers of leaves.

"See," said Tom, "Here are this year's leaves and just below are last year's leaves and below them are the leaves from the year before! I admit that they are a bit soggy and beat-up looking but they are still there."

"Okay," said Laura," But where are the leaves from three years ago and four years ago?"

"I think they are under the ground somewhere," said Tom. "Let's dig some up and see what happened to them."

And so they did and what they found settled the argument.

PURPOSE

My research and that of others show that children have a difficult time understanding the recycling of organic matter in an ecosystem. This story aims to have students speculate about what happens to organic material over time.

RELATED CONCEPTS

- Recycling of matter
- Decomposition
- Closed system
- Decay
- Conservation of matter
- Open system

DON'T BE SURPRISED

Unless students have some understanding of the particulate nature of matter, it is difficult if not impossible for them to understand how matter can be broken down into parts so small that they can be recombined with other matter to form new compounds and impact new living things. Your students may actually believe that leaves and wood from fallen trees over the millennia have accumulated to the extent that they actually increase the mass of the Earth. After all, don't we see little saplings grow into giant trees? Doesn't that add mass to the Earth?

CONTENT BACKGROUND

Obviously, we do not want our students to think that the Earth is accumulating leaves over the millennia and growing more obese. Nor do we want them believing that growing trees and other plants add mass to the Earth each year. It is technically possible that meteorites add a miniscule amount of weight when they enter Earth's atmosphere and finally land on the Earth's surface. True, some hydrogen atoms are lost into space, and rockets and satellites leave our planet for their homes in space, but these are almost negligible. Even though students may bring it up, the amount of mass lost or gained is not worth considering.

We are talking in this story about many, many tons of leaves that have fallen since plants and trees evolved. If you do any walking in the woods in a climate where deciduous trees lose their leaves, you are aware of the depth that these leaves can accumulate. If you have a yard surrounded by trees, the raking and mulching of these leaves takes up a great deal of your effort and time. Yes, the amount of leaves that fall to the ground from trees is substantial. If they stayed in place, I think you can imagine how tall the pile would be by now.

Digging beneath the pile of this year's leaves, you will come across some rather nasty looking wet and bedraggled-looking pieces of leaves, partial ghosts of their former selves. They have already begun the process of being devoured by decomposers

and broken down into their smallest parts, the molecules that made them working leaves, former producers of sugars and starches. There will be numerous animals—mainly invertebrates like earthworms, millipedes, pill bugs—and many varieties of bacteria and fungi at work breaking down the original leaves and plant material into compost, a rich organic soil. They do this by digesting the substance of the organic material to release the building blocks of living organisms into the soil. The farther down you are able to go into the leaf litter, the less the leaves look like leaves and the more they look like soil. These decomposers do a good job, leaving little that is recognizable. Then other living things, including the trees themselves, absorb those building blocks so that they might live and grow. The cycle is complete.

Most of my research and that of my colleagues on the concept of decomposition has shown that children and adults alike believe that dead materials turn directly into soil, as if by magic (Leach, Konicek, and Shapiro 1992; Leach et al. 1992; Sequeira and Freitas 1986). Exceptions to this were students who had some understanding of the particulate nature of matter and believed that it was made up of tiny particles capable of being reused by other living things.

How can all of these trees and plants continue to grow and not add to the mass of the Earth? The answer may be even less believable for your students. Earth is considered a *closed system* since the materials on the planet are finite and stay within the confines of the planet and its atmosphere. As I mentioned previously, virtually no *mass* is exchanged with the other parts of the universe, but of course energy is, particularly the radiant energy from the Sun. When plants make the food and the substances that define their structure and existence, they use carbon dioxide from the atmosphere as their source of carbon. The process of photosynthesis relies on the carbon dioxide in the air and the energy of the Sun to produce the matter we know as living botanical organisms—plants.

But many people believe that the air around us has no weight and no substance. So how can something that has no mass be the source of the carbon that makes up the mass of every living plant? This is a difficult concept for many to understand unless they are convinced otherwise.

When plants decay, the materials are returned to the planet in molecular form, suitable for reuse by other growing things. Thus the mass is conserved over time, so the mass does not change. See the probe "Seedlings in a Jar," which suggests that you place seedlings or seeds in a sealed jar of soil and let the seeds germinate and grow. If the total system is weighed before and after the growth period, it will weigh the same. No matter left or entered the jar so it is a closed system just like the Earth (Keeley, Eberle, and Farrin 2005). See also "Seedlings in a Jar" in *Everyday Life Science Mysteries* (2013). You may want to try this yourself if you are in need of convincing that the closed system will be entirely self-sufficient. Finally, you may also check out chapter 22 in this book, "Rotten Apples," for another take on the decomposition process.

related ideas from the National Science education standards (NrC 1996)

K–4: The Characteristics of Organisms
- Organisms have basic needs.

K–4: Organisms and Environments
- All organisms cause changes in the environment where they live. Some of these changes are detrimental to the organism or to other organisms, whereas others are beneficial.

5–8: Populations and Ecosystems
- Decomposers, primarily bacteria and fungi, are consumers that use waste materials and dead organisms for food.

related ideas from Benchmarks for science Literacy (aaas 1993)

K–2: Flow of Matter and Energy
- Many materials can be recycled and used again, sometimes to different forms.

K–2: Constancy and Change
- Things change in some ways and stay the same in some ways.

3–5: Interdependence of Life
- Insects and various other organisms depend on dead plant and animal material for food.
- Most microorganisms do not cause disease and many are beneficial

3–5: Flow of Matter and Energy
- Some source of energy is needed for all organisms to stay alive and grow.
- Over the whole Earth, organisms are growing, dying, and decaying, and new organisms are being produced by the old ones.

USING THE STORY WITH GRADES K–4

I suggest that you give the probe "Rotting Apple" from *Uncovering Student Ideas In Science, Volume 3: Another 25 Formative Assessment Probes* (Keeley, Eberle, and Dorsey 2008). This will give you an idea of what your students already think about decomposition. Unless I miss my guess, I'll bet that most of your students will choose the option that says wind and water soften the apple and it just dissolves into soil.

With students of this age, it is appropriate for them to observe an apple or some other fruit decompose in a covered container. Making sure the container inside the classroom is closed assures that the fungal spores stay within the container and do not contaminate the room. The presence of obvious fungi will give your students a visual appreciation of their work in decomposing the fruit. In warm weather, it's a good idea to place a container with holes in its lid on the outside windowsill so that it can be observed but not touched. This allows flies to visit the fruit and deposit eggs that hatch and begin to devour the rotting fruit. If maggots or grubs are present, it will give students the idea that larger animals also play a part in decomposition.

Fungi are considered a separate kingdom in the natural world classification order. They get their nutrition by decomposing dead organisms and absorbing the nutrients they need to live and grow. They reproduce by spores that are always present in our air. These spores are the progenitors of the fungi that take advantage of any food left out and exposed to them. They can also be dangerous to some people who are allergic to them. Mold, for example, is a fungus that causes allergic reactions in many people. Entire buildings have been closed because of mold contamination and even condemned and destroyed.

If you do not want to have a decomposing chamber in your classroom, you may avail yourself of the materials on the internet. To view a time-lapse video of fruit and vegetables decomposing, check out *www.metatube.com/en/videos/27174/ Fruit-and-Vegetable-Decomposition-Time-lapse.* Some people find decomposition gross and I think you should view it first before showing it to young children. My personal experience is that viewing how fruit and vegetables decompose through time-lapse technology amazes even young children. Of course, middle school children will find it even more intriguing. You will also notice that the sprouting and growing of certain vegetables in the compost being formed shows the cycle of life in graphic form.

You may want to add a chart of students' "best thinking" on what happens to leaves and other things that fall to the ground. If this is done before letting them look through leaf litter or watching the video of the decomposing materials, they can modify their findings through their experience. (Caution: Leaf litter is a favorite habitat for scorpions, centipedes, and coral snakes if you live in a tropical or subtropical climate. It is best to have your students work with small rakes to find the lower levels of leaves so that they can be a safe distance if they come upon one of these poisonous critters.)

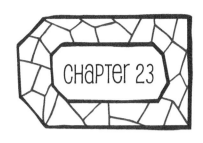

The story in this chapter also goes quite well with the stories and related activities in "Rotten Apples" (Chapter 22 of this book) and "Worms Are for More Than Bait," in *More Everyday Science Mysteries* (Konicek-Moran 2009). All three stories focus on what I consider one of the most important biological concepts—one that is one of the least understood by children and adults alike—decomposition. Understanding that composting (recycling organic material) lessens the burden on landfills promotes stewardship of the Earth. Much of the stuff that is put into our landfills is organic material that could be recycled into usable material. You will find a great deal of information on the Environmental Protection Agency website about what kinds of things can and cannot be composted (*http://epa.gov/recycle/composting.html*).

USING THE STORY WITH GRADES 5–8

Middle school students are more likely to agree with Laura in this story. Giving the probe mentioned in the K–4 section will give you a good look at prior conceptions. Be sure to pay close attention to students' rationalizations in the written section of the probe. It is often in this that you can see if there is real understanding. Only Selma's answer, "I think small organisms use it for energy and building materials," mentions the decomposers. Your students' written explanations will confirm for you whether they really understand the role of decomposers in the process.

If your class ventures outside to recreate the actions of the children in the story, they will certainly notice the differences in the levels of leaves on the ground. (Note the caution about leaf litter containing some dangerous insects and snakes in warmer climates that I made previously.) If students are allowed to dig even farther beneath the soil, they will come upon the macro animals like worms, pill bugs, millipedes, and others that begin the process of decomposing the leaves and other organic material. Help your students to notice the fungi that are also covering the leaves. Looking around, they will notice that the dead trees, either standing or lying on the ground will be covered with fungi. The trees, both standing and fallen often are covered with shelf fungi commonly called "turkey tails" (*Trametes elegans*). These fungi and other plants that grow on the fallen logs help break down the dead trees. Often in moist seasons, trees will take on the role of "nurse logs," as the decomposition of their upper surfaces and the addition of materials falling from the trees above will form a growing platform. They are called nurse logs since they host a great many plants that continue the process of decomposition. These plants are often ferns, but in certain climates, can actually be much larger plants such as trees. The roots of these plants take in the decomposed materials from the nurse log, and thus help to reduce it even further by reusing it to nurture their own growth.

Now let's tackle the concept of plants growing in a closed system and help students understand that plants take carbon from the air in order to make food and build their structure. Give the probe mentioned earlier, "Seedlings in a Jar." This time, use the probe for a discussion starter. Let the students debate, making sure

that they give their reasons why they think the system in the jar will weigh more, less, or the same after the seeds germinate and grow. Then let the students design and set up an investigation using the materials mentioned in the probe. This is the topic featured in the chapter of the same name, "Seedlings in a Jar," in this book. However, be advised that the chapter in this book focuses more on systems. Set up properly, the system will not change weight at any time. Make sure that variables are fully identified and that the investigation is completely fair. It would be a good idea to weigh each part of the system separately: seeds, earth, water, jar, and lid, in order to have these data during the discussion afterward.

Expect that the students might:

- say that the mass for producing the growth of the seedlings came from the soil;
- say that the mass for producing the growth of the seedlings came from the water; or
- accuse someone of adding soil or water or removing some.

Sooner or later the arguments will convince them that whatever happened within the jar used only the matter that was there in the first place. This may not persuade all of your students that the atmosphere contributed the material for the plants to germinate and grow. (The plants will weigh more afterward, after all, and the soil will have not changed significantly, at least not enough to account for the growth of the plants in the jar.) However, it will be another plank in the scaffolding that will sooner or later help them to understand that the atmosphere, indeed, does provide the material that builds the mass of growing plants: carbon from the carbon dioxide.

A similar type of investigation can be carried out in a closed system using decomposers such as worms, bugs, or fungi. An important part of these investigations is that students have the opportunity to plan and carry out investigations and consider the variables involved. Do not use sterilized soil, as it contains no living organisms. You can use leaves or fruit or vegetables to compost. There will be sufficient spores in most dirt to produce fungi. Just remember that if you use animals such as worms or bugs, keep the transparent parts of the jar covered so that the light will not disturb them. The covering can be removed for short intervals in order to allow observation and drawing of what is happening over time.

reLateD BOOKS anD NSTa JOURNaL artICLes

Keeley, P. 2005. *Science curriculum topic study: Bridging the gap between standards and practice.* Thousand Oaks, CA: Corwin Press.

Keeley, P., F. Eberle, and C. Dorsey. 2008. *Uncovering student ideas in science, volume 3: Another 25 formative assessment probes.* Arlington, VA: NSTA Press.

Keeley, P., F. Eberle, and L. Farrin. 2005. *Uncovering student ideas in science, volume 1: 25 formative assessment probes.* Arlington, VA: NSTA Press.

Keeley, P., F. Eberle, and J. Tugel. 2007. *Uncovering student ideas in science, volume 2: 25 more formative assessment probes.* Arlington, VA: NSTA Press.

Keeley, P., and J. Tugel. 2009. *Uncovering student ideas in science, volume 4: 25 new formative assessment probes.* Arlington, VA: NSTA Press.

Konicek-Moran, R. 2008. *Everyday science mysteries: Stories for inquiry-based science teaching.* Arlington, VA: NSTA Press.

Konicek-Moran, R. 2010. *Even more eeveryday science mysteries: Stories for inquiry-based science teaching.* Arlington, VA: NSTA Press.

Konicek-Moran, R. 2013. *Everyday life science mysteries: Stories for inquiry-based science teaching.* Arlington, VA: NSTA Press.

references

Driver, R., A. Squires, P. Rushworth, and V. Wood-Robinson. 1994. *Making sense of secondary science: Research into children's ideas.* New York: Routledge Falmer.

Keeley, P., F. Eberle, and C. Dorsey. 2008. *Uncovering student ideas in science, volume 3: Another 25 formative assessment probes.* Arlington, VA: NSTA Press.

Keeley, P., F. Eberle, and L. Farrin. 2005. *Uncovering student ideas in science, volume 1: 25 formative assessment probes.* Arlington, VA: NSTA Press.

Leach, J., R. Konicek, and B. Shapiro. 1992. The ideas used by British and North American school children to interpret the phenomenon of decay: A cross-cultural study. Paper presented to the annual Meeting of the American Educational Research Association, San Francisco.

Leach, J., R. Driver, P. Scott, and C. Wood-Robinson. 1992. *Progression in conceptual understanding of ecological concepts by pupils aged 5–16.* Leeds, UK: The University of Leeds, Centre for Studies in Science and Mathematics Education.

Time-lapse video of decomposition. *www.metatube.com/en/videos/27174/Fruit-and-Vegetable-Decomposition-Time-lapse/*

Konicek-Moran, R. 2009. Worms are for more than bait. In *More everyday science mysteries: Stories for inquiry-based science teaching,* 91–100. Arlington, VA: NSTA Press.

National Research Council (NRC). 1996. *National science education standards.* Washington, DC: National Academies Press.

Sequeira, M., and M. Freitas. 1986. Death and decomposition of living organisms: Children's alternative frameworks. Paper presented at the 11th Conference of the Association for Teacher Education in Europe (ATEE), Toulouse, France.

INDEX

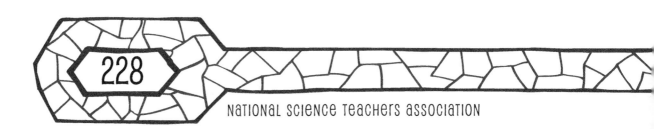

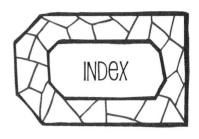

INDEX

S

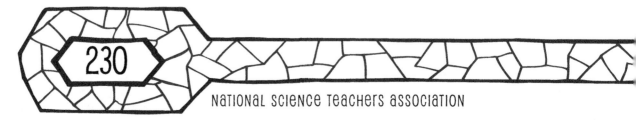

INDEX

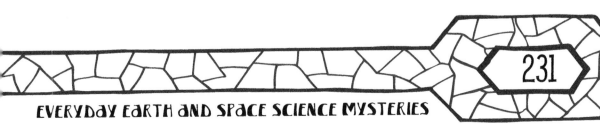

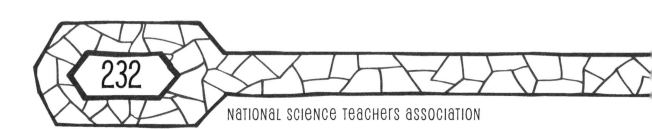